# Science(ish)

## The Peculiar Science Behind the Movies

### Rick Edwards
### Dr Michael Brooks

Atlantic Books
London

First published in hardback in Great Britain in 2017 by
Atlantic Books, an imprint of Atlantic Books Ltd.

This edition published in 2018

10 9 8 7 6 5 4 3 2 1

A CIP catalogue record for this book is available
from the British Library.

Paperback ISBN: 978 1 78649 223 4
E-book ISBN: 978 1 78649 222 7

Printed and bound by CPI Group (UK) Ltd, Croydon, CR0 4YY

Atlantic Books
An imprint of Atlantic Books Ltd
Ormond House
26–27 Boswell Street
London
WC1N 3JZ

www.atlantic-books.co.uk

# Contents

# Introduction

You can pack a lot of hard-hitting truth into a work of fiction. Remember Aesop? The old Greek jackanory? A couple of thousand years ago his fables got some good reviews. Apollonius of Tyana, for example, said Aesop, 'by announcing a story which everyone knows not to be true, told the truth'.

Aesop's greatest hits included 'The Boy Who Cried Wolf', 'The Fox and the Grapes' and 'The Lion and the Mouse'. They all teach us something, making us think about how we should act. And we don't notice, because we're enjoying a diverting story. In other words, Aesop knew how to simultaneously entertain us and make us smarter and better human beings.

The same thing happens when science hits the silver screen. Modern movie-makers are big fans of science. They don't always follow its rules to the letter, but they do see its value to humanity. There is no end of screenplays that show science to be at the core of who we are, what we do, where we're going and what the consequences might be – good and bad. They might be informed speculations, but they are often *very* well informed.

What's more, they invite us to ask some profound questions. Do we need an agency that takes responsibility for diverting asteroids? Is it possible that we could have a global disease pandemic? Can we analyse people's thought patterns, or their shared online data, to predict and prevent crime? Should mogwai be kept as pets?

You might recognize these as movie plots. But it's important to realize that Hollywood doesn't just make this stuff up.* These stories are all based around ideas that real scientists are working on.

The American screenwriter William Goldman famously said that, in Hollywood, nobody knows anything. But Goldman was wrong. Many of Hollywood's directors, producers and writers pay close attention to science. These are clever, creative people, who see what is going on in the scientific sphere and bring it into the light. And so looking at the peculiar science behind the movies is actually a great way to start an important conversation.

In this book you'll be faced with as-yet-unanswerable conundrums about genetic manipulation, the merits of colonizing other planets, creating animals that are part human, the hopes and fears surrounding artificial intelligence, the ethics of de-extinction… There's plenty to think about.

Luckily, there's also some knockabout stuff that probably won't affect the future of humanity. Steel yourselves for the paradoxes of time travel, the mind-bending properties of black holes and the thorny issue of whether we are living in a *Matrix*-style simulation.

We have loved delving into all these questions in our podcast, and now in this book – and we're hoping you're going to have an equally good time exploring these modern fables. Aesop was OK, but we think Hollywood does it better.

---

* Well, apart from the mogwai.

# 1
# The Martian

HOW WILL WE GET TO THE RED PLANET?

IS A MARTIAN HOLIDAY GOOD FOR YOUR HEALTH?

CAN WE REALLY MAKE A LIFE ON MARS?

I love *The Martian*. It's Man vs Wilderness, botanist Mark Watney vs his cosmic fate, Matt Damon vs Ridley Scott leaving him stranded and helpless. It's crammed with science about how humans might live on the surface, what the dusty red soil is made of, what we might be able to grow...

Not sure you'd need an actual botanist for that. Growing plants is hardly rocket science, is it?

Oh really? You think a quantum physicist would be better?

Well, plants are quantum mechanical at heart, with the photosynthesis mechanism transferring energy through the leaf in a superposition state...

Your weird fetish for quantum is embarrassing. The only reason I'd be taking a quantum physicist to Mars is to help the crew sleep through the journey. And as a source of protein.

## Home alone

Ridley Scott's film is based on an excellent and crazily well-researched book (you're right, exactly like this one) by Andrew Weir. While astronauts are pottering around on the Martian surface in 2035, a storm hits. Poor old Matt gets whacked by a broken antenna that pierces his spacesuit and damages the instruments that broadcast his biostats. His friends think he's a goner, so they leave him for dead, blasting off from Mars towards Earth before the storm blows their spaceship over. But this is a Matt Damon movie. So – surprise! – Matt regains consciousness, finds himself alone and with very limited food, and quickly realizes that he's going to have to 'science the shit' out of this situation...

It's a big ask. When you're watching the film, you get the sense that Mars has no mercy. Its dust storms are apocalyptic.

Nothing will grow there. There's precious little water and barely any atmosphere, it's generally nippy by day and needle-sharp cold by night, getting down to minus 125 Celsius in places. Even its reputation is aggro: the colour of the planet Mars, fourth rock from the Sun, reminded the Romans of blood, so they named it after their god of war.

And yet we are ludicrously keen on Mars. The Red Planet has always been an object of fascination to humans, and in the space age never more so. After all, it's not so far away that we can't get there, and although it looks like an alien world now, it was once a bit like Earth. It had an atmosphere, it had water and there's at least some soil you can plant your feet on. If we were to get to Jupiter, we'd find nothing but gas. Jupiter is not a great place to establish a colony. Mars isn't, either, to be honest – it ain't no Center Parcs – but it's a good start.

So the first question that arises is obvious. *The Martian* depends on us being able to get people to Mars. **How are we going to do that?**

## Fantastic voyage

I've just been looking at the Wikipedia entry for Mars One, the colony project. It's hilarious. 'The project's schedule, technical and financial feasibility, and ethics, have been criticized by scientists, engineers and those in the aerospace industry.'

And now it's getting mugged off in our book. Have many people applied?

Amazingly, yes – they had more than 4,000 people pay to apply for places on their Martian holiday camp.

 And are they ever going to get their money's worth?

[Redacted for legal reasons.]

First, you've got to score a seat. Elon Musk, the deep-pocketed founder of SpaceX, says you'll need to pay around $200,000 for a ticket on his flights to Mars – when he's finally ready to issue them. You'll also need to have a 'sense of adventure' and be 'prepared to die'. Well, at least he's honest.

NASA is not currently accepting applications for their programme that will eventually put people on Mars, but it was quite recently. In case they didn't get what they were looking for and reopen the opportunity, here's some of what you need to know.

In the recruitment round that closed in February 2016, the annual salary range was $66,026.00 to $144,566.00. In any event, you'll need a science degree, plus three or more years of professional experience or 1,000 hours pilot-in-command time in jet aircraft. An advanced degree is desirable and you have to be a US citizen. And, would you believe, 'Frequent travel may be required.'

Mars One is the third option. This is also closed to applications just now. But, they say, check back often. Their astronauts must be 'intelligent, creative, psychologically stable and physically healthy'. And without emotional ties. Or financial

## The curse of Mars

In *The Martian*, Matt Damon is left for dead because his fellow crew-members are worried that a dust storm will blow their spacecraft over, stranding them all on the Red Planet. Many people have scoffed at this, because the Martian atmosphere is only 1 per cent as dense as Earth's, and would therefore struggle to blow anything over. However, it has happened before – or at least that's what we think.

The Russian Mars 3 lander touched down on the planet's surface in 1971. It sent a signal home, but the signal was cut off after just twenty seconds. Experts think its mission ended abruptly when a massive dust storm caused the lander to topple over.

Whatever the cause, it's only one of twenty-seven Martian mission failures so far. The problems can usually be pinned on human error, incompetence or inexperience. It started with NASA's 1964 Mariner 3 mission, whose solar panels failed to deploy. Unable to charge its batteries, the craft quickly died. The following year, a solar-panel problem caused the Russian Zond 2 to drift off, lifeless, into space. There was the European Space Agency's Beagle 2 mission, led by the heavily sideburned Colin Pillinger, which landed intact but never called home. There was also the time engineers on the Mars Climate Orbiter mixed up SI and imperial units. Oops.

We're getting a lot better at Mars missions now, though. Most of the failures were last century, and we have run plenty of successful orbiter and lander programmes in the last decade or so. That said, the European Space Agency lost its Schiaparelli lander in October 2016. The curse still has some power, it seems.

commitments on Earth, presumably: there's no pay, as such. Also, the ultimate selection will be by public vote in a TV series, so you'd better have a lot of friends. Or, since it's a one-way ticket, enemies might be more helpful.

Assuming you've got a place, you need to realize it's a long way to the Red Planet. At its closest possible approach to Earth, when Mars is at its nearest to the Sun and Earth is at its furthest from the Sun, it would still be a tiresome 33.9 million miles away. And that has never happened, at least not as far as we know. It would be an inconvenient coincidence, and arguably not worth waiting around for, if the planets were both to hit those points in their orbit at the same time. As they dance around the Sun, the closest these two planets have ever come was in 2003, when they were 34.8 million miles apart. On average, the distance between them is 140 million miles. But there are optimal times to make the journey.

All in all, Mars is a pain to reach. In terms of launch velocity, the fastest spacecraft we have ever built was New Horizons. It initially went to look at Pluto, which is now in its rear-view mirror. New Horizons shot off into the solar system at a pant-soiling 36,000 miles per hour. And yet it would still take around two months to reach Mars. The actual time depends on when you launch your spacecraft at this moving target. As we learn in *The Martian*, there are certain windows of opportunity when a journey is more feasible than others. Working out when these windows will open is complicated but essential, because it takes time to prepare a Mars mission. And it's important to realize that nothing carrying humans will travel as fast as New Horizons, because it would be carrying significantly more weight. New Horizons has little more than a set of fancy cameras. There

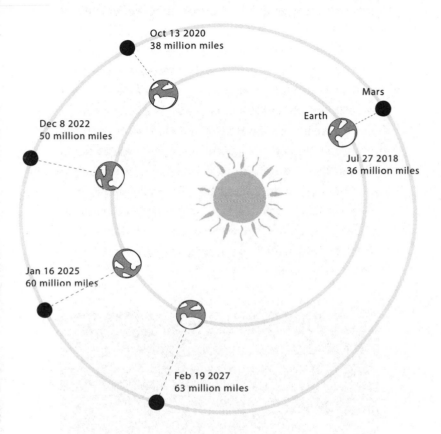

Oct 13 2020
38 million miles

Mars

Earth

Dec 8 2022
50 million miles

Jul 27 2018
36 million miles

Jan 16 2025
60 million miles

Feb 19 2027
63 million miles

**The ideal dates for landing a future mission on Mars.**

are some fast and furious space-travel options coming over the horizon (metaphorically speaking), and we'll get to them when we look into *Alien*. But if you are intent on going to Mars in the next decade or so, you're just going to have to clear a few months in your diary.

## Slingshotting and other cool space-travel tricks

If you want to steer, speed up or slow down a spacecraft, but you don't want to use up fuel unnecessarily, you need to 'slingshot'. This technique, first used by a Russian craft in 1959, uses the gravitational field of a planet or moon and is essential to the plot of *The Martian*. It's actually pretty complicated, but, essentially, if you want to speed up, you do a close fly-by in the direction of the body's movement, gaining energy from its gravitational pull. Fly against the motion, and it's like applying the brakes. You also have to get the angle of approach right for the gravitational field to fling you off in the right direction for your close encounter.

An interplanetary superhighway. © NASA

There are other ways to use the gravity of all the bodies in the solar system. In fact, NASA has laid out an 'interplanetary superhighway' that shows the various possibilities. This is a network of tubes whose walls follow a course set by the gravitational fields of all the various planets and moons.

Put a spacecraft within one of these invisible tubes, and give it a nudge: it will be pulled down the gravitational tube network as if those walls were real, physical guides. Fire a rocket motor at just the right time, and you can move into a different tube at their intersections. However, though it saves on fuel, it's a *very* slow way to get around. If a budget space-flight firm ever offers you a ride on this route, just say no. However cheap the ticket, you'll probably be dead on arrival at anywhere interesting.

Perhaps the most advanced transport plan is the 'Interplanetary Transport System' (ITS). This is the brain-child of Elon Musk's SpaceX corporation, which aims to establish a colony on Mars in the 2020s. Here's how the SpaceX ITS journey works. It starts with a 100-passenger spaceship mounted on a booster rocket that is equipped with a set of 'full flow methane-liquid oxygen' engines. This is powerful enough to lift the spaceship and its 100 passengers into orbit. But it can't carry enough fuel to send them on their way. So, the spaceship-plus-booster combo separates before the ship has reached orbit. The spaceship is put into a 'parking orbit', while the booster returns to the launchpad, lands gently (everyone hopes) and is topped off by a fuel pod

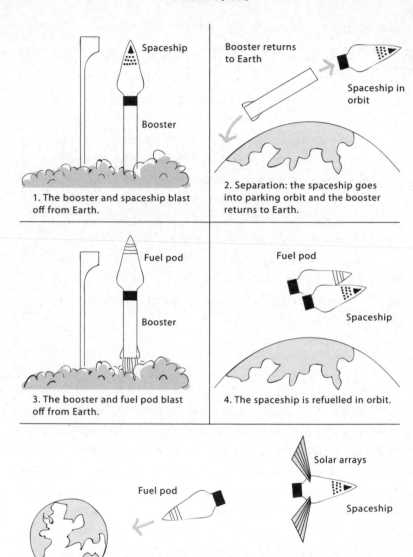

1. The booster and spaceship blast off from Earth.

2. Separation: the spaceship goes into parking orbit and the booster returns to Earth.

3. The booster and fuel pod blast off from Earth.

4. The spaceship is refuelled in orbit.

5. The fuel pod returns to Earth. The spaceship deploys its solar arrays and embarks for Mars.

**Proposed departure sequence for the SpaceX expedition to Mars.**

where the spaceship used to be. The fuelled-up booster rocket returns to orbit, where the fuel pod separates off and fills up the spaceship. The booster and the empty fuel pod return to Earth, and the spaceship is ready to go to Mars. In the SpaceX vision, there will be a fleet of ITS craft travelling together to the Red Planet, so all this has to be done multiple times. Basically, for a short while Earth's orbit will become like a petrol-station forecourt, rammed with Mars-bound ships waiting for clearance to depart. Very cool, Elon.

Months later, at Mars, each ITS descends using rocket-thrusters to perform a controlled landing on the Martian surface. That means the craft will be sitting up, ready for take-off whenever the humans want to go back. Unless it gets blown over in a storm, of course.

Your other option for reaching Mars is a little less enticing. Mars One, a Netherlands-based non-profit organization, is putting together plans for a Mars colony. There is a slight 'budget airline' feel to this operation, largely because you can't buy a return ticket. Mars One describes its Mars Transit Vehicle (MTV) – which is still on the drawing board, by the way – as a 'compact space station' that will carry 800 kilograms of dry food (yum), 700 kilos of oxygen and 3,000 litres of water for the passengers' seven-month journey to Mars. The space station has a separate lander vehicle, which disconnects from the main craft and lands on the red dust, never to fly again. That's right: once you're down, you're down.

The Mars One pamphlet also contains a slightly chilling fact about its bijou little space station: 'The 3,000 litres of water is also used for radiation shielding.' And that raises a second question: **Is going to Mars good for your health?**

# Staying alive

Everyone bangs on about space radiation, but I'm not convinced it's so terrible. On Earth we get about 2.5 mSv every year from the Sun, and from natural radiation in the rocks beneath your feet. An X-ray at the dentist will give you about 0.05 mSv.

 mSv?

Millisieverts. Radiation comes in sieverts, named after radiation-protection pioneer Rolf Maximilian Sievert. Millisieverts are named after his daughter, Milly.

 Ah, science jokes – fun for none of the family.

Before we get onto the problem of the utter boredom of months spent making small talk with a man who won't stop going on about quantum plants, let's talk about one of the biggest problems with space travel: radiation. Space is full of speeding particles often known as cosmic rays. They don't reach Earth's surface because our atmosphere absorbs most of the ones that our magnetic field hasn't already deflected away. However, once you travel beyond Earth's shield, you will encounter plenty of them.

It doesn't have to be disastrous, though. Space scientists have worked out how much radiation passengers are likely to encounter during a trip to Mars by looking at radiation moni-tors aboard the rocket that took the Curiosity rover to the Red

Planet. The result? You'll be within health-and-safety limits, according to Mars One.

Their calculations say you'll get around 380 mSv during the trip. 'This exposure is below the upper limits of accepted standards for an astronaut career,' they say. 'European Space Agency, Russian Space Agency and Canadian Space Agency limit is 1000 mSv; NASA limits are between 600–1200 mSv, depending on sex and age.'

Mars has very little atmosphere and no magnetic field, so colonists will still be exposed to cosmic rays on the surface of the Red Planet. They can expect about 11 mSv per year. That means settlers can work on Mars for about sixty years before they receive the maximum dose that space agencies think is acceptable for an astronaut's career.

We aren't entirely sure, however, that our radiation exposure limits have been set at the right level. Studies have begun to show, for instance, that the Apollo astronauts are suffering unexpectedly high levels of heart disease – possibly because of exposure to radiation that destroyed tissues in their veins and arteries.

Also, the amount of radiation goes up rapidly if the Sun happens to be in one of its active phases, when it sends out 'coronal mass ejections' into space. These huge gobbets of radiation are extremely dangerous, which is why the Mars One craft will have a dedicated radiation shelter – essentially, a huge, hollow water tank – for use when solar-activity forecasts are high.

For the most part, though, the craft's radiation shielding will be enough to keep passengers safe, Mars One says, and you shouldn't expect to spend more than about a week of your journey inside the hollow water tank. Elon Musk, incidentally, doesn't think radiation's a big problem. He's such a space cowboy that he dismisses the health risks that have bothered NASA for decades as 'relatively minor'. As a result, there isn't really much of a shielding plan for the Interplanetary Transport System. Might be worth packing your own lead cagoule.

The physical peril is nothing compared to the mental challenges of life in space, however. First, there's the boredom and isolation. Life on a space station is pretty repetitive, with the same brain-numbing maintenance tasks needing to be done day in, day out. The food is boring. Washing is difficult. It's not a great life.

The selection procedures are designed to root out personalities that are likely to have a problem with this, but they're not perfect, so contingency plans might be necessary. When NASA detects low mood during communications with any of its astronauts, for instance, staff will arrange for treats to be delivered, or for chats with family members. However, on a trip to Mars, this is not really an option. The distance from Earth means that communications are difficult, and it's certainly not possible to send up goody-bags. The postage costs alone would be prohibitive. In theory it's possible that treats could be pre-stashed in secret compartments on the habitation module, to be found by luck or through the transmission of clues, in a weird version of geocaching. However, when people are volunteering, how much responsibility for their well-being do the colonization companies have to take?

Psychologists have studied what might happen during the journey to Mars by confining groups of people together on Earth for extended periods. The results aren't exactly heartening. Studies of people living in Mars simulations have shown that they tend to form into cliques that put the well-being of their own clique members above anybody else's – even if it will jeopardize the whole mission. It's worse if they divide up by gender: men tend to form pacts with each other that prioritize their individual comfort over that of the women. Men are basically dicks.

---

### A typical day on the way to Mars*

06.00    Wake up. Rub down with soapy cloth

06.15    Breakfast – disgusting, as per usual

07.00    Read mission control's briefing for the day

08.00    Housekeeping chores (cleaning, repairs, maybe some ironing)

10.00    Exercise (a losing battle against muscle wastage)

11.00    Snack (dry) and some science research (also dry)

13.00    Lunch (see breakfast)

14.00    Eject rubbish. Weep silently

17.00    Exercise again (giving new meaning to 'star jumps')

18.00    Dinner (see lunch)

19.00    Free time (you can't talk to anyone on Earth any more, so entertain other astronauts with killer anecdotes about when you were a flying ace – again)

19.10    Inexplicably, everyone else has gone to bed early; open up that novel you always meant to read

19.20    Check Facebook and Twitter

19.35    Look out of the window and try to spot Earth – again

20.00    Unpack the rug you stashed in your things, and sing 'A Whole New World' from *Aladdin* – again

20.15    Go to bed; think about killing yourself

* Rick imagines.

Even real astronauts, who are selected and trained to be as mission-focused as possible, can behave badly under the pressures of life in space. In 1973, some of the astronauts on the Skylab space station went on strike for a day because they felt they were being overworked. Then there was the case of the silent cosmonauts: in 1982, two of them went almost seven months on Salyut 7 without talking. Why? They didn't like each other.

If you want to know the other health risks you'll be taking on your trip to Mars, we've compiled a handy list:

## Space flu

Your body did not evolve to cope with microgravity. Your heart is designed to pump against gravity, so on the way to Mars, blood and other fluids will accumulate more in your upper body. The result will be a puffy face, headaches, nasal congestion (in space, everyone will hear you sniff) and skinny little chicken legs. Your diaphragm will float upwards too, making it a little more difficult to breathe. Your back will ache because your vertebrae will float apart without gravity. (On the plus side, you could grow a couple of inches in height.)

## Muscle loss

You'll lose muscle mass because you just don't need to work as hard in microgravity. That means fewer calories are being burned, though. It's lucky the food is going to be so terrible, because if you don't exercise whenever possible, you are going to go to seed. And nobody wants a fat, smelly Martian.

## B.O.

Yes, you will smell. Washing is difficult in space. Not only because a shower is surprisingly gravity-dependent, but because water is a precious resource.

## Nausea

That shift of fluids affects the inner ear, making you nauseous in the first few days. You're very likely to be spacesick. Just under half of all astronauts are, and they've all been chosen because they've got the 'right stuff'. So be prepared to vomit, suffer headaches and dizziness, and generally want to lie down. Except there is no down. Which, as it happens, will also add to your general confusion and disorientation.

## Insomnia

Your sleep patterns are going to change radically. It's often noisy on a spacecraft, and you'll struggle to fall asleep. Your daily sleep/wake cycles are toast, because there is no pattern of darkness and light to give your body the necessary cues. Fatigue is going to hit you like a late-running train. As well as leaving you tired, disoriented and fuzzy-brained, the lack of sleep will also affect your immune system. You're going to catch colds and other viral infections if fellow astronauts are carrying any, and you'll succumb more easily to bacterial infection. Antivirals and antibiotics degrade after a few months, so you'll be mixing your own medicines from dry ingredients. If you're awake enough.

### Bone loss

Eventually, you'll suffer bone loss equivalent to a pensioner, because in microgravity astronauts excrete calcium and phosphorus. That means your bones will fracture more easily, and you might have to pass stones through your urinary tract.

### Psychosis

Psychological effects of the journey include depression, anxiety, insomnia (ha! and you're already so tired!) and, in extreme cases, psychosis.

### Malformed cells

Oh and your cells, especially your blood cells, may not grow and function properly in the long term, because the lack of gravity will change their shape. We don't yet know what the effects of this will be, but come on – it's unlikely to be good.

If, despite all this, you're still intent on a trip to the Red Planet, we need to address our third question: **Can we really make a life on Mars?**

## Life on Mars

So, Brooksy, you and me on Mars. Who's surviving longest?

Let me guess – you reckon that you would?

Well, I'm thinking my general knowledge is going to serve me better than your specialist stuff.

General knowledge? Reading an autocue, you mean?

 You know you're not going to have Google on Mars? It's going to come out that you actually know nothing useful. You're going to die on your arse the minute there's a problem.

At least you'll be there to do the spin-off show. #alwaysthebridesmaid

 I WILL STEAL THE ROCKET AND LEAVE WITHOUT YOU.

GOOD LUCK WITH THAT, ROCKETMAN. I'VE SEEN YOUR DRIVING – YOU CAN BARELY GET HOME UNSCATHED FROM WEST LONDON.

 I'm no psychologist, but I'm not sure that we'd be the best match for a joint mission to Mars...

The crux of the movie is that Mark Watney has to survive four years on Mars before any hope of rescue. If only he knew he was Matt Damon, and thus almost impossible to kill onscreen...

Watney sets out to grow the best food he can, while rationing what's left in the crew habitation tent. His answer is potatoes grown in his own special, personalized brand of fertilizer. You know what we're talking about.

If and when we colonize Mars, however, there are plans to do housing and agriculture properly. The basic Mars habitat (the pros call it a 'hab') is a pressurized tent. It has to be light enough to bring to Mars on a spacecraft, but strong and heavy enough to resist the diabolical Martian weather. It contains a life-support system that includes a breathable atmosphere and heating and cooling. It has to act as a radiation shield and have airlocks that allow the colonists to enter and exit safely. Ideally, it is modular, meaning that you can add and remove sections as they become available or unnecessary. Friends coming over for dinner? Bolt on the conservatory hab.

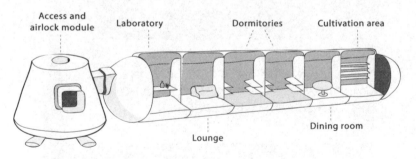

**Possible layout of a Mars 'hab'.**

There are several habs under development at the moment. Much more challenging is the issue of food. The European Space Agency (ESA) has already had a go at planning the vegetable garden, and the menu. It contains rice, onions, tomatoes, soya, potatoes, lettuce, spinach, wheat and spirulina, which is a high-protein algae. You can put spirulina in pretty much anything, and you should: as well as abundant protein, it has a plethora of vitamins and all the essential amino acids (though the taste takes a *lot* of getting used to). ESA has even developed menu-cards (we've improved them, as you can see).

**THE RED PLANET DINER**

**LUNCH MENU**

2 COURSES – $15
3 COURSES – $10

**STARTERS**

LETTUCE SURPRISE* (F)
MARTIAN BREAD & GREEN TOMATO JAM (F)

**MAIN**

SILKWORM & CRICKET STEW (I) (W)
POTATO & TOMATO MILLE FEUILLES (F)

**DESSERT**

SPIRULINA SEMOLINA (F)

All produce is locally sourced where possible.
You'll wish it wasn't. Please let your server know of any food
allergies and intolerances. Then prepare to go hungry.
This menu is not subject to any change, ever.
(I) – contains insects
(W) – contains worms
(F) – grown in the chef's own faeces

*It's just lettuce. Surprise!

**A Martian menu.**

These delicacies will grow in human waste, which will yield water and oxygen as well as nutrients (crapped-out gut bacteria – which don't seem to mind life in space – will be an essential part of maintaining a healthy diet). The plants will need to be grown in greenhouses, because there's no hope of them surviving the sub-zero night-time temperatures or the high levels of ultraviolet radiation that the surface of Mars endures in the daytime. Or, of course, the near-vacuum of the thin Martian atmosphere. There is a plan to take genetically modified, hardened strains of these plants to Mars, but they haven't been developed as yet.

We have already started growing plants in space. International Space Station (ISS) astronauts have, for instance, grown romaine lettuce – and eaten it – in microgravity (it's all carefully contained in closed boxes, to prevent problems like soil floating away). According to those involved, there's something quite special about it. Astronauts say they feel more connected to Earth when they are able to tend plants that have also made their way from our planet. On Earth, salad is only good for your body. In space, it's good for the soul.

Other flowering plants have also bloomed on board the ISS, meaning that it should be possible to put tomatoes on the menu. However, there's still quite a bit to consider. The main concern is that plants evolved for conditions on Earth. Charles Darwin was the first to show that their roots seek 'down', like a plumb line, which is a problem in microgravity. If the roots don't have a defined direction in which to grow, how do you ensure they find the nutrients and water they need?

It's not the only way crops can fail – space gardening is far from straightforward, it turns out. The first crop of ISS lettuce was subject to 'drought stress', which is fancy-speak for 'the astronauts didn't water it enough'. Flowers (the astronauts cultivated zinnia, a kind of daisy) proved even more

## Ultimate survivors

Mark Watney is pretty good, but real-life astronauts have performed some amazing feats, too. In fact astronauts have been science-ing the shit out of space since 1961.

You probably know all about the oxygen-tank explosion aboard Apollo 13, and how its astronauts fixed the damage with make-do-and-mend technology. The lesson there is: if you're heading for Mars or anywhere else in space, take some duct tape. It has been the crucial factor in surviving countless hairy moments.

On the Apollo 11 mission, Buzz Aldrin saved lives with a felt-tip pen. A circuit breaker that was part of the launch mechanism for the Lunar Module had fallen off. That meant he and Neil Armstrong would be stranded on the moon, unless they could fix it. Mission control were scrambling around trying to come up with a fix when Aldrin realized that a felt-tip pen would close the contact, without creating a spark or short-circuiting the mechanism.

Perhaps the greatest display of science-ing the shit out of space came in 1963, when Gordon Cooper's Mercury-Atlas 9 rocket went haywire in orbit. The readouts for altitude, orientation and attitude failed, then the automatic stabilization and the control systems shut down. Then the cabin started filling up with mind-fogging carbon dioxide. Cooper looked out at the stars, worked out his position and orientation, then used his Timex wristwatch to calculate exactly when he needed to fire the retro-rockets to set the craft's attitude and speed for a safe re-entry. In the end, he performed the most accurate splashdown that had ever been achieved. Eat that, Watney!

testing. The watering equipment ended up flooding the roots. The plants had to eject excess water through their leaves, and the poor air flow in the ISS garden meant that they stayed wet. Soon the leaves were covered in mould. Astronauts had to clean the leaves with sanitizing wipes. Then they had to ignore the instructions from NASA and feel their way through caring for the zinnias, just like any gardener on Earth. For all its many talents, NASA mission control doesn't have green fingers, it turns out, and ISS Commander Kelly tweeted about having to channel his 'inner Mark Watney'. However, it all paid off: some zinnia flowers did eventually bloom.

NASA being NASA, an astronaut in space operating independent cultivation is now known as an 'autonomous gardener'. Next up for the ISS autonomous gardener is Chinese cabbage, followed in 2018 by dwarf tomato plants.

Potatoes haven't yet made it onto the ISS agenda – they've only got as far as Peru. That's where the International Potato Center is working with NASA to cultivate potatoes in the Peruvian desert. It's not Mars, but it's not a million miles away from it.*

Why Peru? Potatoes, which are a great source of carbohydrates, protein, vitamin C, iron and zinc, originated in Peru. They are integral to the culture, used in dyes as well as food (and in assessing potential wives: in Peruvian culture, if a woman can peel the particularly lumpy 'weeping bride' potato, she's a keeper). To find potatoes suitable for cultivation on Mars, NASA wants to find varieties that will grow in poor soil in cold, low-pressure, low-water conditions. They're starting with sixty-five varieties, which might produce ten decent crops between them, experts reckon. But

* It is actually.

the stressful conditions could make some of them bitter and inedible.

Of course, it's not just about potatoes. In fact, that would be a disaster. One priority for space chefs is to make sure that the food isn't too boring. Lettuce, tomatoes and potatoes are OK, but not for long. One hi-tech solution to this is a robot chef that is, basically, a 3D printer for food. It sounds futuristic, but the first generation is already with us. It can make cheese-and-tomato pizza from dry ingredients, for example.

So far, there are no plans for pepperoni pizza – in fact, apart from the possibility of eating insects, animal protein isn't on anyone's menu, because it is hugely problematic. We have sent animals into space, but we haven't succeeded in raising any. Quail eggs transported to the Russian space station Mir mostly failed to hatch. The chicks that did hatch displayed developmental defects. Again, the culprit is evolution. All Earth's animals are suited to far more gravity than there is in space or even on Mars. On Earth, an egg's yolk, which contains the developing embryo, sits on the bottom of the shell, which permits the transfer of oxygen between the embryo and the outside world via the porous shell wall. In microgravity, the yolk floats in the middle, and the gas exchange is less efficient, so the chicks are deprived of essential oxygen. Even when chicks did hatch, they couldn't balance in microgravity and couldn't feed themselves. We know this because of experiments carried out on the US space shuttle Discovery, which were paid for by Kentucky Fried Chicken. Maybe Mars, with its one-third of Earth's gravity, would be OK for chickens, but getting them there in one piece remains a challenge.

Hold on. Kentucky Fried Chicken? Really?

Yes. They did two trips in 1989, growing embryos in space on the first one, and then they put some of the offspring of those original embryos into space. The first chick that hatched in space was named 'Kentucky', in case you're interested. They seemed to survive the journey back to Earth OK, and recovered the ability to feed. One researcher was reported as saying that she 'hasn't seen anything out of the ordinary among the space hens and roosters'.

Are Burger King sending pregnant cows into space?

No, Rick. Think about the payload weight. There have been other animal experiments, though. For example, amphibians suffer in space, too, as their instinct is to go 'up' out of the water for air. When there is no 'up', they find themselves in big trouble.

I'm not that keen on eating amphibians, if I'm honest.

What about frog's legs? They even taste like chicken.

Thanks, but I'm all right. I'll eat chickpeas and algae.

Don't be such a lightweight. Anyway, there may be ways to overcome animal-raising in microgravity. We are pretty clever, after all. If we can get ourselves a Mars-friendly potato, maybe we can also breed gravity-independent poultry.

 But it won't come cheep.

Eggsactly.

 [pause] That's enough. So, let's review – Mars. How are we getting there? Hopefully on one of Elon Musk's Interplanetary Transport Systems, so that we can come back when it's awful. Is Mars good for your health? Lord, no. Can we survive? Depends how much you like eating algae grown in your own faeces. *For ever.*

# 2
# Jurassic Park

WERE DINOSAURS REALLY LIKE THAT?

COULD SOMEONE BRING DINOSAURS BACK TO LIFE?

SHOULD WE USE SCIENCE FOR DE-EXTINCTION?

Did you know a doctor called Gideon Mantell was the first person to discover a dinosaur fossil?

I did. The *Iguanodon*.

Very good. Imagine not naming it after yourself, though – that is the very definition of humble. You would name your dinosaur *Rickosaurus*, wouldn't you?

 Obviously. Or the *Rickodon*. Mantell was a fool.

*Jurassic Park* is an undeniably great movie, and one that broke box-office records around the world on its release in 1993. It's hard to believe that anyone doesn't know the premise by now, but here's the lowdown, first laid out in Michael Crichton's 1990 book. Millions of years ago, a mosquito sucks on dinosaur blood, gets trapped in tree resin and is preserved as an amber-encased fossil. Richard Attenborough's character, billionaire John Hammond, funds a biotech programme that extracts the DNA from the mosquito and uses it to bring the dinosaur – many different types of dinosaur, in fact – back to life. He puts them in a theme park, and is almost ready to open to the public. He just wants endorsement from a few scientists...

It's a testimony to the film's quality, and Crichton's prescience, that the movie still feels relatively recent. Perhaps that's because its three sequels have preserved continuity. Part four – *Jurassic World* – came out in 2015 and made clear the enduring appeal of dinosaur resurrection, by setting a new box-office record during its opening weekend. Its dinosaurs are more varied (and better rendered), but still pretty much identical to the ones in the original film. Which brings us to our first question. We have learned a lot about dinosaurs since 1993. So, **were dinosaurs *really* like the ones in *Jurassic Park*?**

# Flight of the *Stegosaurus*

Did you ever want to be a dinosaur-hunter as a kid?

 I still do. Any chance, do you think?

Not now that Xu Xing is on it. He's discovered so many dino fossils he's actually lost count of all the species he's had to name.

 Lucky bastard. What's his secret?

Right place, right time. He never even wanted to be a dino guy. In fact he'd never even heard of dinosaurs until he got to university in Beijing. He wanted to be an economist, but the Chinese authorities told him to do a palaeontology degree.

 Never heard of dinosaurs? That's extremely suspect. I do love it when the state tells you what to do with your life, though. There's not enough of that.

Dinosaurs aren't what they used to be. Palaeontologists have unearthed a bonanza of fossils in China over the last few decades, and this has shed so much light on their physical features that we really have to ditch the dull, ugly reptile stereotype that Gideon Mantell created. Dinosaurs weren't dull. In fact they were probably very striking.

Perhaps the most startling dino-discovery of all is that many of these creatures were distinctly bird-like. But that similarity is not really about flight; it's about being covered in feathers.

The evidence for feathers in dinosaurs comes from fossilized feather-prints and from the discovery of 'quill knobs' (stop sniggering at the back there), which are bumps on the skeleton that help anchor ligaments that support large feathers. These have been found on hundreds of dinosaur fossils – including *Velociraptor* and various herbivorous dinosaurs – excavated from lake deposits in China in the last couple of decades. But we have direct evidence, too. Someone visiting a Burmese amber market, for instance, found a plum-sized lump of solidified resin that contained a dinosaur tail, complete with bones, soft tissue – and feathers. There are other examples of amber that contain dinosaur feathers, along with bird feathers, which were probably blown onto the solidifying resin, where they remained, stuck fast for millions of years.

Did all dinosaurs have feathers? It's a difficult question to answer. In 2014, researchers announced they had discovered the fossilized remains of a small herbivorous dinosaur that had both scales and feathers. This prompted scientists from Sweden, Canada and the UK to create a huge flow-chart that would pinpoint when and where feathers became commonplace. They found that the evidence was too limited to draw firm conclusions, but according to their paper, 'Current data indicate that feathers and their filamentous homologues are probably theropod synapomorphies but fail to support the hypothesis that protofeathers are plesiomorphic for Dinosauria.'

Well, that's cleared that up, then. Now you know why no one wants to sit next to the scientist at a dinner party. Anyway, essentially the answer is: not all dinosaurs were feathered.

## Triassic Period
250 – 200 million years ago

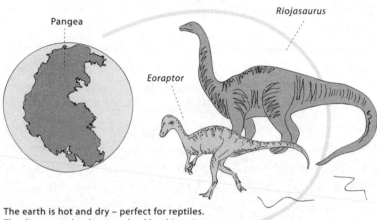

The earth is hot and dry – perfect for reptiles.
The dinosaurs that have evolved by this point
haven't ever become famous.

## Jurassic Period
200 – 145 million years ago

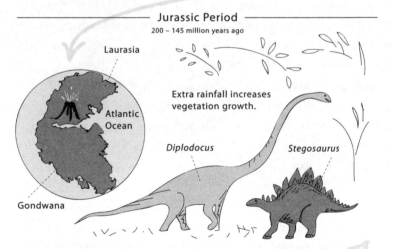

Extra rainfall increases
vegetation growth.

Seismic activity causes Pangea to break up, creating an ocean and two
land masses. Something – no one knows what – wipes out many of the
reptiles and amphibians. The dinosaurs are fine. Still no *T. rex* or
velociraptors, though, so *Jurassic Park* is a questionable naming
decision for a film that revolves around these characters.

## Cretaceous Period
### 145 – 66 million years ago

More continents are appearing, leading to independent evolution of various carnivorous and herbivorous dinosaur species. The list has expanded massively. *T. rex* is leading the cast, with support from *Velociraptor*, *Triceratops* and *Gallimimus*, to name just a few. It's the golden age. But it doesn't last.

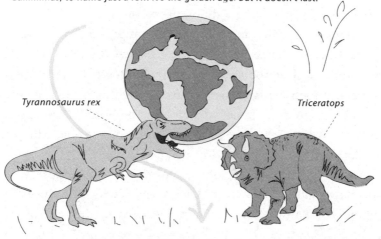

*Tyrannosaurus rex*                               *Triceratops*

## Cretaceous–Tertiary Mass Extinction
### 66 million years ago

10 km-wide ball of rock

Yucatán
Peninsula
(now Mexico)

Unfortunately, Bruce Willis has yet to evolve, so nothing could be done to prevent the asteroid strike. The result includes a big crater and an even bigger dust cloud. For almost all the dinosaurs, it's game over.

**A brief history of dinosaurs.**

Probably. Frustrating as it is to be left in doubt, let's remember that we're talking about unearthing the distinguishing characteristics of something that existed hundreds of millions of years ago. Quite frankly, it's impressive that we've got as far as we have. Even more impressive is the fact that we now know what colour some of these dinosaurs were.

Obviously at this point you're thinking: hang on – we're going from evidence found in the fossil record here. Fossils are mud, infilling the areas where the biological matter used to be and then being turned to stone, effectively. So how the hell are we going to know what colour the original biological material was? It's a good question, and a clever one. But not as clever as the scientists with the answer.

Much of the colour in the animal world comes from specially shaped cells called melanosomes, which produce the pigment melanin. Long, thin melanosomes (they're known as eumelanosomes, but don't worry, there won't be a test on this) tend to produce greys and black. Spherical 'phaeomelanosomes' create an orange-brown pigment. These are the two most common types of melanosome amongst dinosaurs, and so were probably the first to evolve.

Fossilized melanosomes from dinosaur skeletons have been compared against those from zebra-finch feathers and found to be almost identical. This means that mapping the distribution and shape of melanosomes in a dinosaur fossil tells us what the creature actually looked like.

One of the most spectacular applications of this has been to the 125-million-year-old fossil of *Sinosauropteryx* found in Liaoning Province in China. *Sinosauropteryx* was a cousin of *T. rex*, a flightless, meat-eating proto-bird just over a metre long. And, it turns out, the proud possessor of a set of orange and white stripes.

We know this because there are bands down its head, back and tail where there are fossilized phaeomelanosomes, alternating with bands where there are no melanosomes at all. That means its feather colouration would have been alternating orange-brown and white stripes. That's a pretty cool bit of detective work. And it's now been done time and time again. A fossil of the bird *Confuciusornis* from the same period, for instance, says that it was mostly black, with orange tail and wing feathers. Researchers who found a fossilized Late Jurassic dinosaur called *Anchiornis huxleyi* have unearthed so much about the melanosomes that they were able to describe it in astonishing detail. It had a grey and dark body with a red-brown speckled face, white feathers on the long limbs that had 'black spangles', and a rust-coloured crown. See? Palaeontology isn't all digging around in mud and dust.*

Our discoveries about dinosaurs are revealing that they were so visually vibrant that researchers are starting to think it would have influenced their behaviour. So they may have performed courtship displays, like massive-toothed peacocks, and even used their feathers for communication. They might have used them to keep their eggs, or even their hatchlings, warm on a cold Jurassic night: we have found fossils of brooding dinosaurs on nests in what is now Mongolia.

But it wasn't all preening and prancing – far from it. Studies of skulls reveal that many of the avian dinosaurs may have been far more deadly than any bird you care to name. The fossilized skull of *Sinornithosaurus*, for example, has what looks like a space for venom glands, and grooved teeth that would have been able to deliver said venom. The giveaway, say the researchers making this claim, is that the groove runs

* To be fair, it mostly is.

## The softer side of dinosaurs

It might seem a bit odd, given the terrifying stereotype, but dinosaurs seem to have had a nurturing instinct. The evidence comes from dinosaurs that were caught by a lava flow – and subsequently fossilized – while sitting on a nest of eggs (it may have been a brood of hatchlings; it's hard to tell from the charred, fossilized remains). This suggests they were care-giving parents. We would suggest it also tells us they weren't terribly good lookouts. How distracted do you have to be, for the whole family to get fossilized on your watch?

Clearly, getting someone else to do the caring wasn't the solution. In 2004, palaeontologists uncovered a fossilized nest containing thirty baby dinos, plus one older specimen. At first the researchers thought the older skeleton belonged to a parent who had been caught napping by a volcanic eruption. But further examination showed the grown-up wasn't actually old enough to reproduce yet. In other words, it was a babysitter. A babysitter who never got paid – and, you could argue, didn't deserve to.

Being a male dinosaur wasn't all glamour, either: it seems that some had to engage in competitive hole-digging in order to secure a mate. This idea stems from the discovery of deep side-by-side holes that are peppered with theropod tracks (velociraptors belonged to the theropod family). The holes are oval-shaped, about two metres long and forty centimetres deep. They're not the remains of nests, burrows or shelters, the researchers who discovered them reason. Their best explanation is that they are part of a courtship ritual that says, basically, 'I would dig you a really good nest.' The theory might be full of holes, but it's all we've got for now.

from the teeth to that space where the venom glands would have been. In their view (which is admittedly controversial), it's a king cobra with wings. Faced with that bad boy, the feathers would not be your biggest concern.

In fact, *Jurassic Park* remains an accurate representation of just how scary dinosaurs were. Remember that scene where the park's game warden (played by Bob Peck) is hunting down the velociraptors? He gets one in his sights, and she sits, as if waiting to die. Suddenly another beast appears to Peck's left. 'Clever girl,' he concedes, just before his inevitable demise. Setting aside the fact that velociraptors were turkey-sized, not human-sized, it's clear that the *Deinonychus* on which Spielberg's 'velociraptors' were based almost certainly were pack-hunters,

**Which *Deinonychus* is scarier? Maybe Spielberg was right to ignore the evidence for feathers.**

as depicted in the film. We know this from multiple tracks found in China. *Deinonychus* is among the Dromaeosaurs – the 'raptors'. A group of raptors from another species (probably *Utahraptor* – guess where its first fossils were uncovered) left track marks by an ancient river in Shandong Province. The animals were all heading in the same direction at the same time.

And get this. In the movie, Sam Neill's clever-Dick palaeontologist talks about a velociraptor holding its middle toe – the one with the deadly claw – off the ground to keep it sharp. That's exactly the behaviour shown by these Dromaeosaurs: the track marks are around twenty-eight centimetres long, and show only two full toes. The centre one is just a stub, indicating that it has been kept aloft.

However, dinosaurs couldn't spend all their time scaring the bejesus out of everything around them. They had to mate, for instance. And here is where palaeontology becomes the biggest guessing game you've ever played. Examining bones for stress fractures, we can infer fighting behaviours that were possibly to do with mating rituals. But all this guesswork is frustrating. It would be so much easier to work out what dinosaurs were really like if we could just watch them living out their lives. In the absence of a time machine (for now anyway; we haven't got to *Back to the Future* yet), that raises our second question: **Could we bring dinosaurs back to life?**

## Closer than they appear

That Michael Crichton was quite a character: six foot nine tall, married five times...

And a climate sceptic who testified to Congress that global warming wasn't a man-made problem. His take on science was really dodgy.

You're lucky he's dead.

Why?

A political journalist called Michael Crowley would tell you why. He criticized Crichton's anti-science activities. When Crichton's next book came out, he found himself written into the story: as a political journalist called Mick Crowley who had a very small penis.

Wow! That's harsh.

Clearly, Crichton was a sensitive soul.

De-extinction. It sounds cool and futuristic, doesn't it? But it's also possible now. In fact we've kind of done it already, with an extinct wild goat known as the Pyrenean ibex. In 2000, scientists took DNA from Celia, the last living example. Celia died not long afterwards, but in 2003 scientists created a clone. OK, so the clone died almost immediately because of a lung defect, but the proof of principle was there.

Dozens of animal species could be brought back from the dead. There's the dodo, of course, the ivory-billed woodpecker, the woolly rhinoceros, the great auk and the quagga, to name just a few. It's not straightforward, though. We have to ask questions about whether they would encounter a loss of genetic diversity, resulting in inbreeding problems and an inability to adapt to a changing environment, for instance. Can a viable population be brought back? Are there any new diseases that might just wipe them out again? Can it be done – is there enough DNA, and a potential surrogate parent within which the clone can grow?

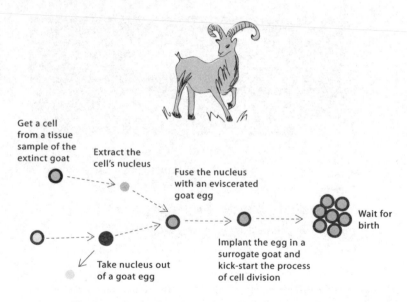

Get a cell from a tissue sample of the extinct goat

Extract the cell's nucleus

Fuse the nucleus with an eviscerated goat egg

Take nucleus out of a goat egg

Implant the egg in a surrogate goat and kick-start the process of cell division

Wait for birth

**How we brought an extinct ibex back to life.**

The passenger pigeon is one species that seems to pass muster. As does the gastric-brooding frog, which (in case you were wondering) keeps its developing young in its stomach and vomits them up when the time is right for birth. This creature went extinct in the 1980s. It would be nice to see that performance again, we're sure you'd agree.

## Pigeon post... mortem

Pigeons haven't gone extinct, as you'll probably have noticed. But the over-hunted species known as the 'North American passenger pigeon' did. Apparently no one noticed that they were gradually becoming less abundant when their numbers dwindled in the twentieth century. The last passenger pigeon, a twenty-nine-year-old called Martha, died at Cincinnati Zoo in September 1914.

That was before the discovery of DNA and before we understood anything, really, about the mechanisms of life. Now we know more – a lot more – and the passenger pigeon has become the flagship de-extinction project.

It began after a group of scientists took DNA samples from the feet of birds that had been stuffed and mounted in various museums. By reading the DNA using a technique called sequencing, then comparing these sequences against DNA sequences taken from pigeons that are still alive and thriving, the scientists worked out that they already had about 75 per cent of the passenger pigeon's DNA. Since DNA is the instruction book for building a new copy of the organism, that felt like enough to work with.

The scientists' plan is that when the technology matures a little further, they will fill in the gaps with DNA from a band-tailed pigeon. Because the speed of reading DNA is rising eightfold every year, and the cost of the required technology is dropping exponentially fast, there is every reason to believe that we can de-extinct the passenger pigeon very soon.

One species that only gets a 'maybe' from the experts is the woolly mammoth. That's interesting, because the experts are pressing on and trying to de-extinct it anyway.

The last mammoth roamed the frozen wastes of Siberia around 4,000 years ago. And it is there that scientists have created what, in a stone-cold rip-off of Michael Crichton's idea, they are calling 'Pleistocene Park'. Pleistocene Park already has bison, horses, moose and reindeer, but that's hardly going to bring the tourist hordes to these frozen wastes. Nobody went to Jurassic Park to see the small herbivores; the public demands spectacle. Which might explain why the owners of Pleistocene Park are so keen to develop their crowning glory: the woolly mammoth.

It is possible – and by two different paths. One is to use DNA from mammoth remains that have been preserved in Arctic ice. This effort, which is led by Japanese cloning experts, plans to inject this DNA into the egg of an elephant, then reinsert the egg into the elephant's womb. Hopefully, the elephant will then grow a mammoth baby and give birth to a species that hasn't been seen on Earth for thousands of years.

Over in the much more pleasant environs of Harvard University, there's a very different plan: George Church and his team plan to build the mammoth DNA for themselves. They know how to do it because we have two complete genomes for woolly mammoths, preserved by the freezing temperatures. One is from around 4,000 years ago, and the other from 45,000 years ago.

The more recent one is from a place called Wrangel Island, which sits in the East Siberian Sea. This is the last known home of the woolly mammoth, and the genome is riddled with the signs of inbreeding – possibly indicating the reason for the species' ultimate demise. The older specimen is the proud owner of admirable genetic diversity, and thus seems like a good one to rebuild. Church and his colleagues can program the DNA sequence into their computers and have chemical-handling robots build the DNA from a few stock chemicals. It won't take long: they are talking about 2018 as the year when viable woolly mammoth DNA will be back on Earth.

The Harvard plan is to splice the relevant, distinctive bits of mammoth DNA into Asian elephant cells, and then use chemical triggers to turn these cells into 'stem cells' that can grow into any kind of tissue. So if you take the hybrid stem-cell nucleus and insert it into the eviscerated egg of an Asian elephant, it should grow into a fat, woolly elephant with enormous tusks.

We know what you're thinking. Yada, yada, yada, pigeon, mammoth, blah-blah, frog, dodo – WHERE'S MY VELOCIRAPTOR?

Fine, let's deal with the dinosaurs. Are they next in line for resurrection? It would be nice to be able to say yes. But all the experts say no. The problem is: DNA degrades, rotting away like a leaf in your grandad's compost bin. The dinosaurs lived so long ago it seems impossible that viable DNA could still be found.

According to killjoy Morten Allentoft of Denmark's Natural History Museum, it takes just over 500 years for half the molecules in a strand of DNA to have decomposed into something else. So when you're dealing with things that were living 65 million–230 million years ago, we're talking about a lot of rotting. After a few hundred thousand halvings, there's not much left, even when you started with all the millions of letters of DNA.

Even worse, somebody has actually checked out the whole 'DNA from insect entombed in amber' thing. David

Penney, who works at the University of Manchester, is an amber expert. He was always suspicious about claims made in the 1990s that DNA had been extracted from a mosquito that had been trapped in resin, then fossilized. So he enlisted the help of a DNA expert called Terry Brown, and picked out some choice specimens of insects in amber (they're not that hard to find; you can get them at a jeweller's). They picked relatively young samples, up to 10,000 years old, and extracted the DNA from the trapped insects. It turned out the DNA had degraded just as badly – worse, actually – as the DNA in the ancient air-dried insects found in many museums.

We're sorry, but it's just not scientifically possible to bring them back. Crichton was lying to us all along. But, to soften the blow, we can offer you: dino-chickens.

Dinosaurs are being re-created, just not in the way you might expect. If the work of Anjan Bhullar and Arhat Abzhanov of Yale University pans out, the Jurassic Park of the future will be populated by chickens with velociraptor snouts. It's perhaps not going to be a massive draw, but you'd pay a bit of cash to see that, wouldn't you?

How are we doing this? Well, let's not forget that birds are dinosaurs. Their ancestors were around in Jurassic times, and there's a clear and obvious link when you look at them. Imagine if birds were bigger: they'd be terrifying monsters. And they'd be even more terrifying with a velociraptor snout.

Of course, these scientists' work is not officially about making dino-chickens. Officially it's about discovering the molecular pathways that turned a snout into a beak as the dinosaur population evolved into birds. But turning back the clock on 150 million years of evolution clearly tickles the fancy of some biologists.

Dinosaurs had two bones that supported the snout (modern-day reptiles still have them). In birds, these bones grew and fused together to form the beak. That seemed to happen because of the action of two proteins. Scientists have blocked these proteins in chicken embryos, which then grew with snouts rather than beaks. Scans of their skulls looked uncannily like fossilized *Archaeopteryx* and, in some cases, *Velociraptor*.

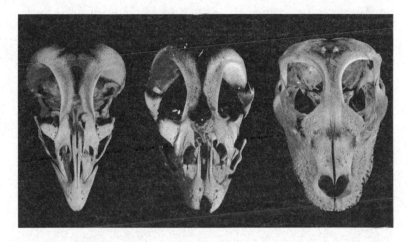

The dino-chicken. These are the skulls of (L–R) a normal chicken, the dino-chicken, and for comparison, the dinosaur-related alligator. © Anjan Bhullar

Cool as that undeniably is, there are other ways to play this game. You can put dinosaur tails on chickens by modifying the chicken genome, for instance. It's quite tricky, though: there are lots of genes that might be involved, and finding the right ones to tweak is difficult. But genetic modification has at least given us chickens with dino leg-bones. Modern birds have short, tapered fibulae that are more like splinters and don't even connect with the bird's ankle. But not the genetically enhanced chickens in Alexander Vargas's lab at the University of Chile. His student Joâo Botelho has learned how to inhibit a gene called Indian Hedgehog (if you want more great gene names, the *Gattaca* chapter has a list), and this causes chickens to grow a long, dino-esque fibula that is properly connected at the ankle. OK, it's not *T. rex*. But you have to start somewhere.

Or do you? In fact, should you? It's something scientists are not always great at asking. So we'll make that our final question: **What are the ethics of de-extinction?**

## Back from the dead

Favourite character?

The only one with a sense of ethics: Jeff Goldblum's mathematician. He has that killer line: 'Your scientists were so preoccupied with whether or not they could, they didn't stop to think if they should.'

Ah, Ian Malcolm. Two first names, like all the best people. Matt Damon, Michael Douglas, James Dean, Katy Perry, Fiona Bruce, Bill Murray, Rick Edward(s)...

Ron Jeremy.

I don't know who that is.

'The lack of humility before Nature that's being displayed here staggers me.'

Ian Malcolm has a point. At first glance, bringing back long-dead species looks like a good idea. After all, if we have the technology to bring back species – especially species that humans have driven to extinction through hunting or habitat-destruction – we almost have a responsibility to use it, don't we? Nobody stopped the killing of the passenger pigeon, did they? And if we can bring it back, then that doesn't matter, does it? All's well that ends well, as Shakespeare put it.

Not necessarily. This is exactly the kind of thinking that makes the case for those who say we *shouldn't* begin to de-extinct. In short, they claim that de-extinction removes the sense of urgency and peril from conservation efforts. If you suggest that you can repopulate from DNA samples, no one is going to worry about conservation. And people *should* worry about conservation.

In the last 500 years human activity has forced more than 800 species into extinction. Species are going extinct at a rate that is unprecedented since the demise of the dinosaurs. More than 40,000 species are currently on the International Union for Conservation of Nature's 'Red List', which denotes a threatened state. More than 16,000 are threatened with imminent extinction. Most primate species are now on the endangered list, as are one in four mammals, one in eight birds and one-third of all amphibians. Add to that the fact that around 70 per cent of the world's plants are also in danger of extinction, and you can see how the problem is going to escalate if we don't do something.

So how can we defend taking our eyes off the conservation ball and focusing our effort on de-extinction? Especially when, as with the woolly mammoth, de-extinction will involve use of the wombs of some endangered species (in this case, the Asian elephant). Perhaps, for a start, because we can do it intelligently. By, for instance, not going for the glamour animals, but for the ones that have most chance of fitting into the existing ecosystem.

That means swapping the velociraptor project for ones involving the Réunion giant tortoise or the lesser stick-nest rat, which recently disappeared from Australia's bush. Why? Because Nature is a complex web; to quote *The Lion King*, it's the Circle of Life. That means considering habitats, prey and predation. If you've got somewhere suitable to live, something good to eat and can re-create a broken link in the food chain – in other words, it was just bad luck or human folly that drove you extinct – your resurrection as a species is going to be a good thing.

That's good news for the Christmas Island pipistrelle bat. No one knows exactly why its numbers declined rapidly during the 1990s, but it is now widely believed to have gone the way of the dodo. That's a problem, because it was the only bat in the neighbourhood that would eat insects. After this species disappeared, there were a lot more insects around, and nobody (except the insects) wants that.

You might be wondering how the Réunion giant tortoise qualifies. After all, it can't be eaten, and it only ate plants, so it didn't control any animal populations. Well, it dispersed plant seeds through its poop. And since the tortoise's demise, those tortoise-poop-dependent plants are also heading towards extinction. You see? The Circle of Life demands the return of the pooping tortoise.

The other qualification for resurrection is simple: can we do it properly, creating an abundance of animals that might have some genetic diversity, some ecosystem impact, and go on to re-establish a sustained population?

It's a question that all of the various organizations trying to conserve species are asking in different ways. The International Union for Conservation of Nature, for instance, has issued an official set of guidelines on resurrecting extinct species, and has pointed out that there are plenty of reasons for caution beyond the simple conservation argument. It could introduce species that become dangerously invasive, for example, wiping out existing species (we're looking at you, Australian cane toad). It might provide new ways for diseases to spread, or unwittingly revive ancient bacteria and viruses that we can't control. It might resurrect species that destroy crops and livelihoods, and even kill people. In all, the IUCN lists five advantages and twelve disadvantages of de-extinction. If you go by these numbers, the implication is clear.

One important consideration, when talking about resurrecting ancient species, is how well they would live alongside humans. Would we be able to keep them away from cities, or safe from poachers, trappers and hunters? How much is mammoth-tusk ivory going to be worth? How long would it be before some bellend with a high-calibre rifle pays a few million dollars for the chance to shoot a resurrected sabre-toothed tiger?

What we don't want is for de-extinction to be the first step on the path to re-extinction. That would achieve nothing, and would almost certainly halt any large-scale resurrection project in its tracks. So we have to do this carefully, or not at all.

We also have to think about what gets resurrected, or created for the first time, when we bring an ancient animal back to life. If we have ancient species roaming the Earth again, any number of parasites will take advantage of this new niche and almost certainly adapt to maximize the opportunity. As we've mentioned, humans will also, potentially, have

## Chaos theory — will resurrection always go wrong?

Chaos theory 'simply deals with unpredictability in complex systems'. That's what Jeff Goldblum tells Laura Dern. Of course that wasn't enough for Laura, forcing Jeff to turn up the flirting to eleven, play with her hair, hold her hand and talk about the imperfections in her skin creating a vastly different set of possible outcomes for the water drop that falls on her fingers.

In truth, it's not so much a theory as a statement of what happens in certain situations. In these situations there exist physical characteristics or properties whose behaviour is extremely sensitive to a change in conditions. And this may have been what did for the dinosaurs. Why? Because the orbits of the planets and various asteroids in our solar system are another example of chaos. The planetary orbits are generally stable, but the orbits of the rocks in the asteroid belt are highly sensitive to small changes in the gravitational pulls around them. A particular alignment of the planets, which might occur once every 100 million years, can be enough to shift an asteroid out of a stable orbit and onto a trajectory where new gravitational pulls combine to throw it way off its original course. Now its orbit is chaotic, and anything can happen. Including hitting Earth. According to researchers at the University of California, Los Angeles, there was a period of chaotic disruption roughly sixty-five million years ago: about when the dinosaurs went extinct.

It's not just about physics. The natural world seems to be a chaotic system too. Its chaos means that you can get catastrophic ecosystem collapse as a result of a tiny change in the number of insects, for example. Or rendering an animal infertile will trigger an ability to clone itself, as happens with some species of shark. Or slivers of spliced-in genetic code can create instabilities that will change the fundamental behaviours of a creature such as a velociraptor. In other words, we mess with Nature at our peril.

a range of new viruses and bacteria to contend with. Given that we're already facing an antibiotic crisis, with evolution throwing out life-threatening bacteria that can't be killed by any medicines in our possession (just wait till you get to the chapter on *28 Days Later*, if you're think we're over-egging the problem), do we really want to create a breeding ground for a new set of biological enemies?

And yet... would you really say, 'No, don't do it'?

It's a much more complicated, nuanced issue than we might have thought. Kudos to two-first-names Ian Malcolm for calling it. So, to sum up...

The dinosaur world is not like Jurassic Park. It is beset by feathers, preening, prancing, hole-digging and poor babysitting. We can't bring them back, and some people are saying we shouldn't even try anyway.

You sound angry.

I'm fuming. I want to see dinosaurs. I want to go to Jurassic Park. I want to be halted in my tracks as a herd of *Gallimimus* careens across my path through the countryside.

You're being a baby. Put a pipistrelle bat on your Christmas list and get over it.

's first encounter with Hollywood,
th the wormhole space-travel concept
an's book and movie about extrater-
And Thorne was actually a character
king biopic *The Theory of Everything*. He
enzo Cilenti (who is, by the way, in *The*

in a future when the Earth is becoming
rming is increasingly difficult because of
kind of pervasive crop disease, and human-
home. Unfortunately, some short-sighted
hballed NASA many decades ago, and so
pe of that – OR IS THERE?
series of frankly unbelievable occurrences,
A flying ace Joseph 'Coop' Cooper (Matthew
ey turned up to eleven – and not in a good way)
at a few plucky individuals have been keeping
of travelling to the stars alive with a secret space
e. Cue all sorts of crazy plans to use a black hole
me handy rips in space and time, provided by ben-
liens – as a portal to a better future.
ight need to do something like this one day. Plenty
ntists think our only long-term hope is to colonize
rs, and we might need the help of a black hole. So
es sense to ask the most obvious question first: **Are**
**holes real?**

# 3

# Interstellar

ARE BLACK HOLES REAL?

WHAT HAPPENS IF YOU FALL INTO A BLACK HOLE?

DO WE REALLY NEED QUANTUM DATA?

This film is very special. It was written in collaboration with super-scientist Kip Thorne. He's like a god to me, one of our greatest living scientists. He did his PhD under John Wheeler, who invented the term 'black hole'. Thanks to his work, *Interstellar* features the first realistic depiction of a black hole.

No it doesn't. In 1979, French astrophysicist Jean-Pierre Luminet used a punch-card computer to work out what one would look like. He didn't have a printer, so he drew the result of his computations by hand – and it looks quite like *Interstellar's* black hole, Gargantua.

How do *you* know *that*?

You mean, when *you* don't? Because, unlike you, I read Kip Thorne's book about the film. Ever heard the phrase 'fake fan'?

*I*nterstellar is not just a blockbuster, it's a scientific sensation. The man who dreamed it up, Kip Thorne, is a brilliant astrophysicist and one of the geniuses behind the 2016 detection of gravitational waves, whose existence was predicted by Albert Einstein 100 years previously.

Thorne wrote the original screenplay for *Interstellar* (which Steven Spielberg was going to direct) and gained a credit as Executive Producer. It wasn't all one-way traffic, though. Thorne and his colleagues used the enormous amount of computer power available to Hollywood's CGI behemoths to do some new scientific calculations on the nature of black holes. These insights were subsequently published in the peer-reviewed scientific literature, taking movie-science to a whole new level. *Interstellar* has literally given us a new scientific view of what a black hole is like.

In
time.
at the
out to E
at 99.8 pe
according
gravitational
a stable orbit. 1
more slowly on
intense gravitation
It means that for eve
on Earth.

What about the stun
this is the most impress
because it wasn't created
by the science. Everyone sta
falling into the hole, illumina
like a disc – it's known as an 'ac
scientists worked out exactly w
found that the black hole's warping
would actually warp our view of th
program crunching the numbers pre
weird halo effect, where the accretion d
below and in front of the black hole. At firs
was just a computer error. And then the sc
that it was the unexpected truth about wha
would look like.

This wasn't Thorne
either. He came up wi
for *Contact*, Carl Sa
restrial intelligence.
in the Stephen Haw
was played by Vin
*Martian*).
*Interstellar* is se
uninhabitable. F
some unspecified
ity needs a ne
politicians mo
there's little h
Through
former NAS
McConaugh
discovers t
the dream
programm
– plus so
evolent a
We r
of scie
the sta
it ma
**blac**

# A hole in space and time

So here's a slight problem I had with the basis of the film. The alien beings live in a black hole and have the technology to open up a wormhole that offers a shortcut through the fifth dimension.

What's your point?

If you had that kind of technology, presumably fixing the crop-blight problem would be a piece of cake. Couldn't they just deliver a big can of hyper-dimensional Super-Weedol and go back to the fifth dimension?

Are you saying Kip might have over-engineered the solution to what is, essentially, an agricultural hiccup?

I worry that your mate Kip thinks big physics always has the answer.

People like you will be left behind when we move planet, you know that?

Perhaps the biggest and best-developed character in *Interstellar* is the black hole, Gargantua. It is this awe-inspiring object that, according to the movie, offers humanity's only hope of survival.

It's a lot of pressure on what is, in many ways, a very delicate concept. Black holes have had a difficult time over the years. These days almost everyone has at least heard of them, but there was a time when the most eminent scientists wanted them to disappear.

It was an Indian mathematician called Subrahmanyan Chandrasekhar – Chandra for short – who first took black holes seriously. He was performing some calculations about what happened to stars at the end of their life, and noticed that if they were heavy enough, they would collapse under their own weight. To understand why (and what a black hole really is) we have to understand a bit of Einstein's general theory of relativity. Don't worry, it's not that hard.*

Einstein's theory is an upgrade to Sir Isaac Newton's theory of gravity. Newton's theory described how one object moves under the influence of another object's mass. It enabled him to calculate the orbits of the planets, which pull on each other because of their mass.

Einstein went one better and described *why* these objects move as they do. It starts with the idea that space and time are not fixed, flat arenas for our existence. Instead, they are distorted by mass and energy in the same way that your mass, and the energy of you jumping up and down, will distort a trampoline. This distortion puts a curvature into the space and time (commonly just called spacetime) around any massive or energetic object. This means that something that happens to be trying to travel in a straight line through this bent space is actually going to follow a curve. So gravity, which looks like something being pulled towards something else, is actually about being deflected

---

* Well, actually, it is hard. But we're going to go easy on you. And us.

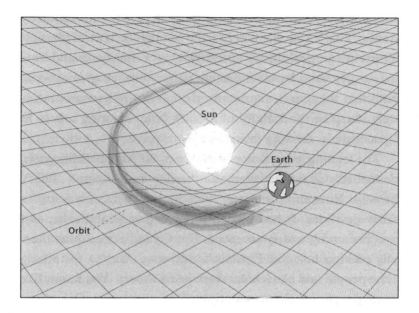

Gravity is simply a distortion of space and time. Because the sun has mass, it distorts the space around it, causing Earth to 'fall' towards it.

off your intended course, because your path through the universe is curved.

Let's go back to Chandra's idea. A star is just a ball of burning gas. As it burns, it creates an outward pressure that pushes against its own gravity and keeps it pumped up. But once the fuel is all used up, there is nothing but the atoms and molecules that are created in the fireball. The mass of each of these atoms and molecules creates a gravitational pull on all the others, shrinking the dying star. As it gets smaller, that gravitational attraction gets stronger, and the star gets even smaller and denser – and so on, and so on. If the star was big enough in the first place, the end result of this is that

produce. And the wobble that LIGO saw was an exact match. So, no, we haven't seen a black hole, exactly. But, thanks to LIGO, we're now damn sure they're real.

Black holes rotate around each other and then merge.

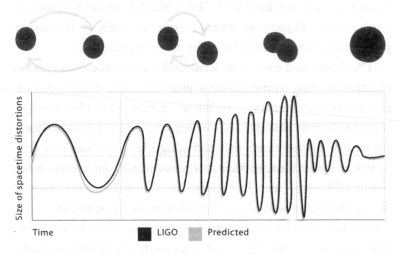

**LIGO's sighting of a black hole merger. The grey line is the gravitational wave signal we would have expected to detect with LIGO, our most sensitive detector. The black line is what we actually detected. The match between them is almost perfect.**

And if black holes are real, then we can – in theory – explore them, even if we can't, like Matthew McConaughey, use them to exploit the hyper-dimensional nature of true love. (We're as disappointed as you are, frankly. But trust us on this. Outside of the Disneyverse, you can't rely on True Love's Kiss.) However, caution is required. Black holes are not to be messed with, as you'll find out from our second question: **What happens if you fall into a black hole?**

# Strung out

What did you make of the line that love is 'the one thing that transcends time and space', that it's observable and powerful, that it has to mean something?

I thought Coop's response was perfect. Its 'meaning' is its utility in social bonding and child-rearing.

That's ice-cold. Do you tell your wife how much you value her social utility? I can't imagine being like that.

That's because you haven't been married as long as I have.

When Coop falls into a black hole he discovers the meaning of love. It's unlikely you'll get the same enlightenment. But we can't be sure what you will discover.

For such a simple question, this one has a *really* complicated answer. Or should we say 'answers', because there are quite a few possibilities, including time travel and adventures in a parallel universe. There are even issues of perspective, because what happens depends on who the 'you' is. If you're the one watching, the observed outcome is very different from the experience of the one who's doing the falling.

In the movie, Matty Mac (as we believe he likes to be known) crosses the event horizon of the black hole. Once past this point of no return, he has no way of ever getting out again

– IN THEORY – but somehow (no spoilers) – IN PRACTICE – he does. We're not going to criticize this evasion of natural consequences unduly, because it came from the god-like mind of Kip Thorne, and who are we, frankly, to question his decisions? However, it's a far cry from what we think will happen.

Let's start as we approach the event horizon. No, on second thoughts, not 'we'. We'll watch from a safe distance. You. You're going in. Feet first. Because we want it to be the most fun possible.

Ahead of you is a glorious blackness, the most complete dark you've ever seen. By the time you approach the event horizon, your feet, which are approximately two metres closer to the singularity at the heart of the black hole, are experiencing a much stronger gravitational pull than your head. And so you are being stretched by what are known as tidal forces. Physicists – the fun ones, at least – call it 'spaghettification'. You're being spaghettified, made long and thin, by a black hole.

Let's assume that you're falling into what's known as a supermassive black hole, a really big one like Sagittarius A*. This is the black hole at the centre of our galaxy. Sagittarius A* has an enormous gravitational field, and because of some fairly complicated physics, the difference in pull over two metres at the event horizon of Sagittarius A* doesn't stretch you to breaking point. If, on the other hand, you were falling into one of the punier black holes, your head would be pulled off before you even crossed the event horizon. And where's the fun in that?

The fun doesn't stop with spaghettification. Oh no. For a start, you're now travelling in time, not space. The intense gravitational field beyond the event horizon causes space and time to bend so much that they actually swap roles. So you

## Other black-hole movies are available...

Warning: some of these should be spaghettified.

### The Black Hole (1979)

The crew of the USS Palomino discover a spaceship parked next to a black hole. How is it not sucked in? It generates a mysterious 'null gravity' bubble. Inside the black hole, we discover later, live weird creatures. Maybe that's where Kip Thorne got the (terrible) idea for the beings in the fifth dimension...

### Lost in Space (1998)

It's the year 2058, and pollution is making Earth uninhabitable. (A bit like a catastrophic crop blight, isn't it, Kip? We're seeing a pattern here...) In this movie, which stars Joey from *Friends*, the black hole is created by the collapse of a planet. Which, as we know, isn't a thing in any universe where the normal laws of physics apply.

### The Black Hole (2006)

Judd Nelson of *The Breakfast Club* grows up into a particle physicist whose atom-smasher accidentally opens up a black hole. From which a weird creature emerges. Maybe it was this one, Kip?

### Sphere (1998)

We're on the floor of the Pacific Ocean, exploring a seemingly alien spacecraft that is believed to have got there via a black hole. Actually, it's a US spacecraft from the future. The plot, which comes from a Michael Crichton book, goes downhill from there. Dustin Hoffman, Sharon Stone and Samuel L. Jackson have all done better work.

---

**Treasure Planet** (2002)

A not-terrible *Treasure Island*-in-space animated movie in which Emma Thompson voices a ship's captain who looks a lot like a cat. The black hole is a bit incidental here. No physicists were harmed – or even consulted – during the making of this movie, we're guessing.

---

are now moving through time, which not even the best of our technology can control. The endpoint of your journey – the singularity – is now as unavoidable as tomorrow. It is, effectively, a moment in your future, not a location in space.

Weirdly, you are oblivious to all this distortion. Because you are now part of the whole thing, it all looks normal to you. To us, looking from the outside, nothing could be further from the truth.

Imagine we're parked a safe distance from the event horizon. The intense gravitational field we are looking at does weird things to the light that does manage to make it back to us. As you were falling towards the event horizon, the light reflected off you was stretched by gravity, acquiring a longer and longer wavelength. To us, you are turning red.

As if that weren't enough, the intense gravitational field slows time, and you appear to be falling in ever-slower motion, never quite reaching the event horizon and disappearing from view. In other words, we could watch your inevitable, reddish demise for ever. What a treat.

Anyway, back to you and your enviable experience. Here comes the singularity! And what happens next is, quite frankly, informed speculation. Some people say you'll simply be crushed to death by the gravitational forces. Jollier physicists say that the singularity causes a new bit of spacetime to form, and you'll enter what is effectively a new universe. Like we said: fun!

Also fun is the idea that you'll emerge in a different part of our universe because the black hole is actually a wormhole – a portal – between different parts of space and time. That means, as we'll discover in *Back to the Future*, that a black hole is potentially a means of time travel.

Some physicists have even suggested that falling through a black hole's singularity is the way to access some 'extra' dimensions of space, breaking out of the boring three spatial dimensions you've experienced all your life and finally taking that well-deserved holiday in the fifth dimension (time is the fourth dimension, in case you thought we'd lost one somewhere). What it's definitely not, though, is a way for Matty Mac to appear somewhat creepily behind the bookshelf in his daughter's bedroom. Sorry if that's a spoiler, but really, it is *very* unsettling and it's best you know about it now.

One more thing. Everything we've just told you might well be wrong. Why? Because Einstein's general theory of relativity is definitely wrong. Yes, that's right: Einstein didn't have all the answers.

To be fair to him, he did give us a good start. But his prediction of gravitational waves is, in some ways, his undoing. The fact that we've seen them caused by black holes means that black holes are real. And if black holes are real, but general relativity can't actually describe what happens at the infinite curvature of the singularity, that means there's something missing from the theory. It's incomplete. It needs some help. It's going to be replaced by a better theory that can do the job properly. Go to the bottom of the class, Albert. No, in fact go to the head's office and bring back the quantum data.

*Interstellar* is peppered with references to this 'quantum data'. It's the key to everything: human survival, understanding black holes, navigating the universe, extracting a pot of hummus without ripping its cardboard sleeve... OK, maybe not the last one, but definitely the others. So that has to be our third and final question: **Why, exactly, do we need the quantum data?**

## A solace in quantum

One of the things I really liked about the movie was that the robots weren't humanoid. When you think about it, it makes sense. You wouldn't bond with them in the same way. So it would be much easier to jettison them into a black hole, say.

It was refreshing, wasn't it? I also liked that you could alter their settings for truthfulness, humour and trust. I wish I could alter some of my friends' settings.

But would you want them to alter yours?

No need. Mine are perfect.

Truthfulness might need a tweak. And we might as well turn your humour setting up from zero.

Unless you've got the memory of a goldfish, you'll remember that general relativity is not up to the job of describing everything in the universe. For that, we'll need what physicists like to call – somewhat unimaginatively – a theory of everything.

Imagine, if you will, a beam of light travelling across the universe from V762 Cas, the most distant star we can see with the naked eye. Relativity describes the path light will take past all the intervening planets and stars, with their gravitationally induced cloaks of warped space. Quantum mechanics, on the other hand, describes what happens when a single photon of that light finally arrives at your eye after a 16,000-year journey and interacts with a single molecule in the retina of your eye.

We don't have a single theory that can describe how that photon interacted with the gravitational fields it encountered on the way. That's because quantum mechanics and relativity are utterly incompatible. Physicists don't know how to make relativity, our best description of the universe on cosmic scales, mesh with quantum mechanics, our theory-in-chief of the very small. That matters, because it's the only way we'll fully understand how the universe began.

The Theory of Everything (ToE) that physicists are seeking will need to be based on 'quantum gravity', this elusive and as-yet-unwritten marriage of relativity with quantum theory. And our best chance of creating a quantum-gravity theory is to get to grips with what is going on inside black holes, because gravity makes them form, while quantum mechanics describes the infinitely small point at the heart of any black hole. Black holes are literally the place where quantum meets gravity.

It seems the key to all this, though, is not at the heart of a black hole, but on its edge – the event horizon.

## Wanted: a Theory of Everything

The quantum data, when we find it, could bring us some surprises. You know about atoms being made of electrons, protons and neutrons. Maybe you know about protons and neutrons being made up of little particles called quarks. But what's the next level down in the nature of reality?

We just don't know. At the moment, our best guess is that everything – whether it's matter or energy – is ultimately composed of vibrating loops of energy. Physicists call them 'strings', and have created something called 'string theory' to describe how they act to create the reality we're familiar with.

String theory is just a mathematical idea at the moment. It has no experimental backing, and very little prospect of any during our lifetimes. But it does at least make interesting suggestions.

One is that the various different subatomic particles arise from the different ways in which the energy strings vibrate. Another is that there must be lots of unseen dimensions to space – seven or eight of them, depending on the exact flavour of string theory that you're looking at.

Where are these hidden dimensions? Various experiments have tried to find them, to no avail. That's no surprise, say the string theorists: they're all around us, but rolled up into long, thin tubes that are so narrow we just can't detect them. This is called 'compactification' (nowhere near as good as 'spaghettification'). It might be a good call, or it might be the most elegant fudge in physics – and there's a lot of competition for that title.

String theory is not the only attempt to unify relativity and quantum mechanics into a quantum theory of gravity. There are other options, such as 'Loop Quantum Gravity', 'Causal Dynamical Triangulation' and 'Twistor Theory'. Like string theory, all of them are almost certainly wrong.

To understand why, we need to take a look at a quantum phenomenon called the Uncertainty Principle. This says that there are limits to the defined properties of anything following quantum rules (which, in a universe ultimately composed of matter and energy, is everything). Essentially, you can't know everything about anything: you can't know a particle's position without losing knowledge about its momentum, for example.

The energy of empty space is one such thing that can't have a precise value. The Uncertainty Principle says that, over a short interval of time, you simply can't know the energy that's held in a volume of empty space. And if you can't know it, that means it can't be zero. Not exactly. The result of this is that there is always a bit of energy coming and going, no matter how 'empty' a chosen volume of space might appear.

According to quantum theory, the not-exactly-zero energy of the universe actually manifests as little pairs of 'virtual' particles that spontaneously pop into existence. The pairs are matter and antimatter and, when they meet, they annihilate.

In 1974, Stephen Hawking pointed out something extraordinary about this. If the particle and antiparticle appeared at the event horizon of a black hole, it might be that one fell into the black hole and the other didn't. Then they couldn't meet and annihilate, and there would be a new, extra particle of energy in the universe, spawned out of the black hole. This creation of energy costs the black hole some mass, Hawking noted. That's because Einstein's relativity tells us that energy and mass are interchangeable (you know this from the formula $E = mc^2$, with $E$ as energy, $m$ as mass and $c$ being the speed of light). So the black hole would constantly be losing mass, and eventually it would have none left. It would cease to be. It would be an ex-black hole. It would have evaporated.

Black-hole evaporation through Hawking radiation has a very odd consequence. Not only does it mean the black hole disappears from the universe, but it also means the information about everything that has ever fallen into the black hole also disappears. But a cast-iron law of quantum theory is that information is a fundamental part of the universe and you can never destroy it.

There are various ways we might resolve all this. The obvious one is to say that the information comes out in the Hawking radiation. Physicists offer various arguments about why this can't happen – and they're pretty good arguments, which is why we have been left with what's called the Black Hole Information Paradox.

Theoretical physicists have been trying to solve the paradox for forty years now, and things have got pretty wild. There is nothing quite like the mind of a theoretical physicist for a source of truly weird ideas. And there's nothing quite like a black hole for stimulating the weirdness. The latest solution involves a spherical shell of fire that incinerates anything falling past the event horizon before it can get into the region from which information would be lost.

However, this 'black-hole firewall' creates its own problems. That's because relativity says someone falling under gravity into a black hole shouldn't notice any strange things happening to them just because they're crossing the event horizon. And it would be hard not to notice that you're on fire. Even if you'd had a drink or two to steady your nerves.

Is there a way out of this? Not yet, but there are some even crazier ideas. One involves a barrier made from matter in a frozen quantum state – a sort of particle-based wall of ice, in other words. Another solution is that the black hole never forms properly: instead, the collapsing star 'bounces' up again like a rapidly inflated balloon at the last minute. Another suggestion is that time flows backwards inside the black hole, causing information to flow back out. We only know one thing for sure: none of these are right.

There is a more prosaic solution, and it's one that we might at least have a chance of testing. What if the information never actually falls into the black hole, but stays on the surface of the event horizon, trapped at the boundary where time and space swap roles? If the information is sitting there, maybe we can see how it is encoded, giving us a huge clue to how gravity and the quantum intertwine. In other words, we'd have access to the quantum data.

Amazingly, theorists are starting to think about how to skim off this quantum data (if it's there). At the moment, their best hope is through finding something in the details of gravitational waves. A gravitational wave that arises from the merger of two black holes, for example, might have a shape that is related to the quantum data on the holes' event horizons. If that sounds like a long shot, it is. But unless someone is willing and able to go inside the black hole, emerge in another universe (maybe) and somehow get the quantum data back to us (unlikely), it's the best hope we've got...

I've enjoyed this more than I enjoyed the film, if I'm honest. These are BIG questions, aren't they? I'd love to be around long enough to see quantum gravity worked out.

I know what you mean. If there was one thing I could do for humanity, it would be to create that final theory, and explain the Big Bang.

I think we'd all be grateful if you created any theory. Or, indeed, anything useful. Anyway, to recap: black holes are real, you really don't want to fall into one, but if you did, you might – just might – end up in another universe...

And we definitely do need the quantum data, in case anyone fancies scoping out the interior of Sagittarius A*.

I've just had a thought. Maybe there is one thing you could do for humanity, Michael...

# 4

# Planet of the Apes

. . . . . . . . . . . . . . . . . . . . . . . . . . . . . . . . . . . . . . . . . . . . . . . . .

HOW DID HUMANS END UP ON TOP?

COULD ANOTHER ANIMAL EVER TAKE OVER?

CAN WE ENGINEER SUPER-SMART CHIMPS?

. . . . . . . . . . . . . . . . . . . . . . . . . . . . . . . . . . . . . . . . . . . . . . . . .

Before we get stuck in, can I just check something: are we talking about the 1968 original or the 2001 remake with Marky Mark Wahlberg? Or the new ones – *Rise of...* and *Dawn of...*?

Well, I started rewatching the original with Charlton I-Love-Guns Heston and felt my brain beginning to devolve. It's absolutely dreadful.

For once we are in total agreement. And Tim Burton's isn't a lot better.

 Maybe that's why he said he'd rather jump out of a window than do a follow-up?

Luckily, no one asked him to. Because I think I'd rather jump out of a window than watch it.

 OK, let's just do them all then. One good thing is that they've made me think about potential prequels to this book. *Rise of the Science(ish), Dawn of the Science(ish)*…

Who's Dawn? Is she replacing me?

 If only.

For all its faults, this is a classic of the 'what-if' genre. What if humans weren't the dominant species? What if we were treated by apes the way we treat them? What if other animals could talk? What if Tim Burton had said: No, someone else could do a better job?

Without ever explicitly stating it, the film is about the trials and tribulations of evolution. The biggest mistake people tend to make when thinking about evolution is to believe that there is an ultimate aim. Evolution has no purpose. In simple terms, random changes in an organism's DNA sometimes give rise to new features. These new features might be useful; more often than not, they are useless. Occasionally they are worse than useless, and make survival less likely. The best features survive because they positively affect the organism's relationship with its environment. This is natural selection. So in this massive, long-running game of chance, here's our first question: **How did humans end up on top?**

## Out of Africa

What do we mean by 'on top'? I reckon that if bacteria could talk or think, they would consider themselves to be the dominant group.

Because of sheer weight of numbers?

Yes. Earth is home to ten million trillion times as many microbes as humans. There are about sixty trillion single-celled organisms in your body.

It's not really my body then, is it? My own cells are outnumbered.

On the plus side, all that weight gain might not be your fault.

We'd better start with two disclaimers. First, 'on top' is subjective, of course. We're not the most numerous species, but we are the most impactful, and the least threatened by other species. We're not saying that's a good thing, just that we are the dominant force on the planet – the only ones with the power to screw it up for absolutely everything else, for example... (see *Jurassic Park*).

Second, we're mainly relying on a patchy fossil record, which means that much of what we think we know about the evolution of humans, and their route to the top, relies on intuition and inference rather than clear, incontrovertible evidence. In fact evidence is very thin on (and under) the ground. So it's nearly impossible to prove that any given factor was solely responsible for an evolutionary change. Often it's classic chicken-and-egg stuff: did we get big brains in order to sustain large social groups, or did we get big brains and then start to live in large social groups? Did we get smaller teeth because we stopped biting each other, or did we stop biting each other because our teeth got smaller? Actually it was more complicated than that, but – spoiler alert – we'll probably never know the whole story. So, given all that, this is what we *think* led to human dominance...

Twenty million years ago or so, you couldn't move for bloody apes. There were at least a hundred species knocking around. But then, thanks to changes in the world's climate, the big blocks of forest started getting smaller: bad news for our ape friends, who were well adapted to the woodland life. With less of their favoured habitat available, lots of ape species went extinct. Our mob survived (clearly) and around seven million years ago said 'See you later, losers' to the common ancestor of us and our closest living relatives, the chimpanzees. No specific fossil remains of that ancestor have been found, incidentally, but we have confidently inferred its existence from other fossil evidence. Darwin thought it must have been 'a hairy, tailed quadruped, probably arboreal in its habits'.

In all that time since the split, we haven't diverged much genetically – the genome of chimps, like the ones who are ruling the roost in *Planet of the Apes*, is 98.5 per cent identical to that of humans.* There are some remarkable similarities. We have the same number of hairs per square inch as great apes, we have the same blood types and some of their behaviours will feel familiar – chimps display aggression, support, betrayal, sexual politics, grief, self-awareness and cultural practices that are different in different groups.

One common misapprehension about us as a species is that we are somehow more evolved than the other primates. It's not true: we've simply been evolving on a different branch of the family tree. Chimps and other apes are like our hairy cousins, and have also been evolving.

We don't know the exact shape of our 'branch' of the family tree. What we do know is that it wasn't simply a straight line, with one ancestor evolving into the next and then the next, until we reach modern-day humans. There have been lots of lineages branching off, meaning that many of these distinct species, our relatives, were living at the same time. All but one of these lineages – that of *Homo*, which emerged almost three million years ago – are now extinct. Not only that, but we are the only surviving *Homos*. We definitely won that particular evolutionary competition.

---

* Don't get too excited – about 50 per cent of our DNA is the same as a banana's.

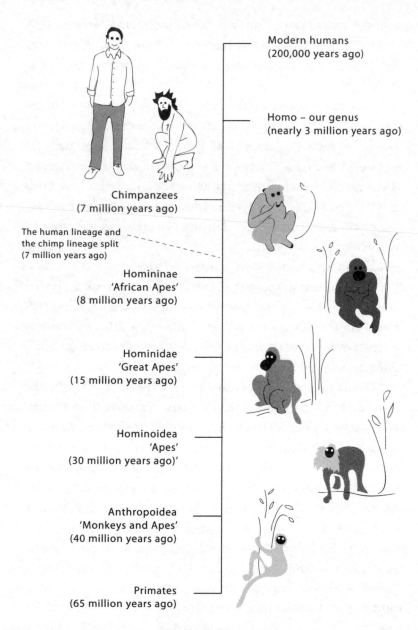

Modern humans
(200,000 years ago)

Homo – our genus
(nearly 3 million years ago)

Chimpanzees
(7 million years ago)

The human lineage and
the chimp lineage split
(7 million years ago)

Homininae
'African Apes'
(8 million years ago)

Hominidae
'Great Apes'
(15 million years ago)

Hominoidea
'Apes'
(30 million years ago)'

Anthropoidea
'Monkeys and Apes'
(40 million years ago)

Primates
(65 million years ago)

**Our family history.**

Having parted ways with the chimps, it doesn't seem that a great deal happened for the next few million years. Yes, our group, the hominins, was evolving, but they remained, in essence, furry little apes with small brains, long arms and big teeth, hanging out in forests. They did start walking on two legs – bipedalism – and venturing out into the savannah to look for food, but there was little sign of what was to come. You wouldn't have looked at them and thought, 'Just you wait, those guys'll get to the moon.'

Then, suddenly, they started to get their act together. The order of things is, as ever, unclear. But we know that a number of developments were happening. You'd guess that loping around on two legs would leave your hands free, and three million years ago our ancestors started leaving examples of simple stone tools. These tools may have paved the way for our brain growth. Once you have sharp tools, you don't need sharp fingernails and your rudimentary manual dexterity improves. Once you can start cutting and mashing up your food, you don't need such strong jaw muscles or big teeth to chew with. That might seem inconsequential, but it may have been the key in us getting a big ol' brain – the main difference between us and our chimp cousins.

On average, an adult chimpanzee brain weighs 384 grams. An adult human brain, by contrast, weighs almost a kilogram more: 1,352 grams. It's fairly safe to assume that a big brain was a major part of how we ended up on top.

A significant factor in the expansion of our brains was a single, lucky mutation in a gene called *MHY16*. In general, primates have strong jaw muscles, which effectively keep a tight grip on the cranium and restrict its growth. If your skull can't get bigger, neither can your brain. That's Brain Growth 101. But this *MHY16* mutation caused our jaw muscle to be made of a different protein and thus become smaller, allowing the skull and brain to start expanding. This might have been an unfavourable development, had it not been for the development of stone tools to help with our food-processing. It's extraordinary to think that had this mutation not occurred in our ancestors, but in, say, the ancestors of another great ape, this book might be getting written by a couple of orang-utans. Not that you would know the difference.

It's worth pointing out that evolving a big brain is not all good news for a biped. For a start, there's the birthing problem: bipedalism requires a narrower pelvis, which restricts the size of the skull that can pop out without fatally injuring the mother. Evolution has got round this by humans giving birth to underdeveloped young whose craniums continue to grow after birth. Human brains grow four times as much as chimp brains in their first two years of life.

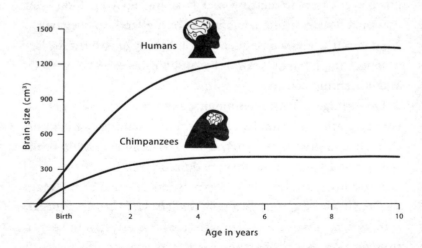

Being bipedal requires a narrow pelvis, which means giving birth to small-skulled babies whose brains grow rapidly after birth.

Then there's the energy issue: a big brain is extremely expensive to run. A modern human brain makes up just 2 per cent of our body mass, but uses 25 per cent of the energy. That's a problem because upright walking was a factor in shortening our digestive tract, which makes it harder to extract energy from our food. To keep our ravenous brain running, we would have to be eating raw food solidly for more than nine hours a day and that's nigh-on impossible, not to mention boring.

We solved this problem by cooking. Raw foods require a lot of chewing and digestive work, in order to be broken down for energy. But around one million years ago we started to control fire. Shortly after that, we were channelling a prehistoric Jamie Oliver and cooking up a storm. The process of cooking breaks down food into easily absorbed sugars. It effectively acts like an external stomach that gives us a virtual extension to our shortened digestive tract. Similarly, we're not missing our strong jaws any more because we don't have to chew everything as much.

Cooking changed a lot of things. One fringe benefit was that the fire kept predators away. So instead of hauling ourselves into the trees at night, we could stay at ground level without fear of getting attacked. With more energy per mouthful, we no longer needed to spend every waking hour eating. That freed up time to do other stuff – like forming social bonds that made life easier by enabling a division of labour. In particular we were able to share child-caring duties, allowing us to have more kids in less time – handy if you're looking to evolve. It also meant we could specialize: in root-gathering, tool-making or putting our well-fuelled big brains to use in innovative forms of hunting, for example.

At some stage we started hunting using snazzy new tools. Projectile weapons – throwing bits of sharpened stone, basically – meant we could take down some big animals and eat their meat and bone marrow. This will have helped provide the nutrients and energy required for further brain growth. Having good weapons at your disposal also means that you can kill other hominins – even someone bigger and stronger than you. This may have levelled the playing field somewhat and encouraged groups of our ancestors to play nice and get along. Large social groups will have emerged, and our big brains would have been able to keep track of who everyone was, what they might be thinking and whose side they would take in an argument. That's important, because you didn't want to find yourself excluded in a world where loners would be fire-less, shelter-less, food-less and easy prey for the savannah's sabre-toothed cats, hyenas and the like. Exclusion from the group would have meant almost certain death.

Another helpful thing about a big brain with lots of processing power is that it provides the means for a complex spoken language to emerge. Obviously language doesn't fossilize (that's the annoying thing about behaviours – they will not turn to stone), so we are stuck with educated guesses about the era of its origin. However, other primates have a balloon-like appendage on their larynx that enables them to make loud, intimidating booming sounds. We seem to have lost that at least 600,000 years ago, and that loss might be one of the developments that enabled us to form more distinct sounds and therefore words. We also have a specific variant of a gene called *FOXP2*, which appears to help coordinate the complex motor mechanics of speech. That's believed to have been around for more than 550,000 years. Obviously the physical capacity to form words doesn't necessarily imply that anyone was having in-depth conversations – they may just have been singing around the campfire, as a form of bonding. That being said, sophisticated tool-making and group hunting activities would have required at least some form of basic communication. So the best guess for the origin of language is a rather imprecise one: between 1.6 million and 600,000 years ago.

However and whenever language came about, it's obviously an essential part of the human story. Without it, society as we know it simply could not exist. Prior to language, evolution and environment controlled our destiny, and cultural change was highly limited. Information was passed down from generation to generation in our genome. But with language, we could share vast amounts of knowledge as we chose. We began, instead of adapting to our environment, to adapt the environment to us. And then we passed on that information to the younger members, who could continue to make use of the experience of their elders. This created a society that was ultimately able to survive whilst becoming almost exempt from the pressures of natural selection. And that is unique in the animal kingdom.

So that's how we got there: with tools, cooking, language and, above all, the big information-processor in our head. But are we going to stay at the top? Here comes our second question: **Could we be displaced by another species?**

## One of life's losers

Here's another 'what if'. The 2001 movie got through an amazing roster of potential directors before Burton ran with it. Alan Rifkin, Peter Jackson, Oliver Stone, Chris Columbus, Roland Emmerich, Sam Raimi and Michael Bay were all linked with the script. What if any of them had taken it on?

Well, for starters, Tim Burton and Helena Bonham-Carter might never have met, and we wouldn't have suffered the travesty that was *Sweeney Todd*.

They might have met elsewhere.

But then he wouldn't have seen her in her 'cute ape' prosthetics. I think that was probably what swung it.

One of the extraordinary features of Tim Burton's *Planet of the Apes* is the primates' ability to leap as if the ground is a trampoline beneath their feet. Every time there's a battle, chimps are bouncing into combat with a forty-foot jump through the jungle.

Clearly, chimps aren't really able to do that. But next to humans, they are pretty athletic. In fact, compared with a whole swathe of animals, we are pathetic creatures. We aren't particularly strong, or fast. We have no body armour, and no lethal weapons at our fingertips. Naked, unarmed human vs a hungry lion, an angry gorilla or a boa constrictor in any kind of mood is not a fair fight. Surely these animals could outcompete us, wipe us out and take our ecological niche?

Well, that depends on whether they can compete with our intelligence and our technology. Other species definitely have the ability to use very rudimentary 'technology'. Tool-use, for instance, is well documented amongst chimps, which have been observed sharpening sticks with their teeth and then using them to spear bushbabies. Some dolphins use sponges to protect their beaks when foraging on the seabed. New Caledonian crows can fashion tools out of leaves or sticks to get at their food.

Our other big advantage, lost in the original *Planet of the Apes*, is language. After all, that's what has enabled us to spread knowledge, teaching each other – and the next generation – vital survival skills.

Again, the roots of this advantage exist in many species. Lots of animals communicate with each other. Whales sing, bees dance, dolphins apparently give each other names (and gossip behind each other's backs). We've taught chimps sign language, and squid communicate using colour and pattern. Vervet monkeys give different warning cries, depending on the type of predator that's approaching.

However, most researchers in the field wouldn't call any of this 'language', exactly. It's certainly not in our language league: there is a richness to our communication that enables us to pass on complex information to our peers, and to our children. In other animals, that ability is nowhere near as developed, as far as we can tell. And that is almost certainly why humans have become so numerous, with a global distribution, shaping (or trampling upon) the environment as we please.

If the hierarchy is to shift then, there would have to be a big disruption. There are various natural possibilities, but the most likely threat comes from within: all-out nuclear war or an engineered supervirus that escapes from a research lab. (That last one is more likely than you might expect, as you'll find out in our look at *28 Days Later*.)

So, let's take the first option, where humans knock themselves off their perch in a nuclear war that wipes out all the large mammals. What happens next? Who steps up to rule the roost?

# Things that could wipe us out

We're on top now, but this is no time to be complacent.

## Sun expansion consumes Earth

**Timescale:** five billion years

**Threat level:** 11 – certain annihilation of everything

**Solution:** set course for another star's planetary system, pronto

## Global nuclear war

**Timescale:** always worryingly imminent

**Threat level:** 8 – some living things (and probably humans) will survive

**Solution:** start lead-lining your cellar

## Asteroid strike

**Timescale:** impossible to say, but we're probably OK for the next century

**Threat level:** 9 – ask the dinosaurs

**Solution:** Bruce Willis

## Nasty viral pandemic

**Timescale:** any time

**Threat level:** 7 – a really nasty cold. That kills millions

**Solution:** live in an isolation chamber and give up all human contact

**Simulation turned off**

**Timescale**: could happen at any moment, sorry (see *The Matrix*)

**Threat level**: if we're living in a simulation, 10; if we aren't, 0

**Solution**: stop talking about living in a simulation – it might piss them off

**Artificial intelligence takes over**

**Timescale**: it's already happening. Maybe (see *Ex Machina*)

**Threat level**: 6 – we can turn them off

**Solution**: keep them busy playing Go and old Atari games.

Although we know that dolphins and whales have exhibited great intelligence, and they would probably survive (and thrive) in the wake of a humanity-destroying event, it's tempting to rule out all the water-dwelling creatures. For all their sponge-on-beak innovation, dolphins don't have anything approaching hands. Could they realistically manipulate their own environment? Moreover, they aren't in a position to start mastering fire, for obvious reasons. And fire is important. Not only because it enables the release of more energy from food to get that brain growing. Further down the line, unless they want to be stuck in a second Stone Age, the dolphins will want to start smelting, to make metal tools and Regency-style garden furniture.

That said, there is one sea-critter worth a little side-flutter: the octopus. These animals seem pretty smart. They can problem-solve and manipulate objects with astonishing dexterity. They can open jars with their tentacles. They can build shelters underwater. Some octopuses can also move around on land. So who's to say that, given enough time and opportunity, they won't all come slithering out of the water and grab some kindling?

However, if, rather than a nuclear war, a species-specific virus that infects only humans took us out, it does seem as if the chimps are the best candidate to replace us in our niche. They are our closest living relative and therefore by definition have the least far to go. Of course there is no guarantee that chimps would move into our niche, if they did survive. And even if they did move in, we simply don't know if they would ever evolve human-like intelligence. If they did, it would still take millions of years. And that's barring disasters: a supervolcano or another massive asteroid strike wouldn't just interrupt the evolution of a new intelligent species – most likely, it would wipe it out. So that would be another reset.

Of course it's entirely possible that in the post-apocalyptic scenario, no intelligent life gets to rule the roost. We don't know that human-level intelligence is an inevitable conse-quence of evolution. After all, the dinosaurs ruled the planet for 160 million years and there's little to suggest that intel-ligence was key to their evolutionary success.

Would the world be a strange place without an intelligent creature? Not at all. In fact, you could argue the contrary: the world is currently a strange place because there *is* an intel-ligent species dominating. Before modern humans took over, 200,000 years ago, no single species had ever dominated. For millions of years there were different apex predators in different locations: a diversity of ecosystems and a diversity of animals, none lording it over everything like humans do. So there is a clear post-human scenario where the Earth and its creatures breathe a sigh of relief and revert back to that.

## Rise of the rats

In his book *The Ancestor's Tale*, Richard Dawkins considered a post-nuclear world in which all large mammals are wiped out. That means no chimps, so who takes over? The rats, maybe. Like the tiny mammals that survived the asteroid impact that took out the dinosaurs, rats are small and can find places to hide away. They are the ultimate post-human scavengers.

Of course no animal is pre-adapted to a future environment, meaning that the rats' adaptation will occur very, very slowly. However, it can be handy to be a creature that reproduces at a furious pace: the mutation rate in the genome is higher, and therefore favourable adaptations will emerge more quickly. That puts them at the front of the race to exploit the niche that has just blown wide open. Given that and their decidedly unfussy eating habits, their population would explode.

These post-apocalyptic halcyon days wouldn't last for ever. Eventually the abundant food would run out and the rats would start turning on one another. But intense competition for survival and quick turnover of generations are another great combination for evolution. On top of that, populations would be isolated from one another again – rats aren't stowing away on ships any more – so they'd be evolving in independent groups. That is likely to give rise to evolutionary divergence into available niches.

What would happen? Well, they might not all stay small. Rodents have form for getting huge: three million years ago there was a monster rodent, *Josephoartigasia monesi*, which weighed about a tonne. Where larger species vanish, smaller species will take advantage. So there might be massive grazing rats being preyed upon by big-toothed predator rats. Glorious stuff. Perhaps even a species of intelligent rats would emerge, one that gives rise to rodent historians and scientists who attempt to 'reconstruct the peculiar and temporarily tragic circumstances that gave ratkind its big break', as Dawkins puts it.

One other possibility worth considering is that *Homo sapiens* speciates. Evolution is still occurring within humans: we can track the mutating genes, which are also mixing more than ever because of the unprecedented flow of people travelling around the world. We don't seem to be speciating, though, because there are no natural-selection pressures. We have adapted our environment to us, not the other way round. Technology allows us to thrive without significant genetic change.* But what if a new, highly unusual niche were to come up?

Something that seems likely, from our look at *The Martian*, is that we will colonize the Red Planet. That raises a very interesting possibility. The human population on Mars will be largely isolated from the one on Earth. The environment on Mars will, of course, be quite different – with much weaker gravity, for starters, and more exposure to gene-mutating radiation. In other words, the conditions for that human population to speciate into a new species are pretty good. Mars might be our galactic Galapagos. Perhaps, thousands of years later, this new *Homo* species would then come back to the mother planet, where boring old *Homo sapiens* is living

---

* Of course we might just merge with machines, which would be a pretty big change.

out its Earthbound days. The newcomers would presumably be looking to occupy an almost identical niche and might well outcompete us, eradicating our kind.

If you think that couldn't happen, look at the history of the Neanderthals. Or the Denisovans. Or any of the early humans that we shared the planet with and eventually drove out of existence. Ever wondered why it is that we feel so distinct from the rest of the animal kingdom? Maybe – and it's still only a maybe – it's because we killed off all of our closest relatives.

There is one more way that non-humans could usurp us. For all our apparent cleverness and ruthless self-preservation, it's possible we might be displaced by something biological of our own creation (or non-biological – see *Ex Machina*). Medical researchers looking to find cures for human diseases have a new set of tools that could close the gap between us and our closest living relative, the chimpanzee. So, could the scenario in *Rise of the Planet of the Apes* be brought into reality by accident? That's our third question: **Could we engineer super-smart chimps?**

## Monkey see, monkey do

Why haven't we spoken about the possibility of plants taking over? Maybe it should be *Planet of the Grapes*?

Very good. First of all, plant photosynthesis is all very clever, but the energy yield is pretty low, so they can't develop brains. Second, they don't have the energy to move around easily enough to be a threat.

What about carnivorous plants? They get energy the same way as some of the most successful animals do. And they don't need roots, because they get water and nutrients from the insects they eat. So they could evolve brains and the ability to move around.

So *Little Shop of Horrors* is possible?

I was thinking *Day of the Triffids*. Which says a lot about our respective personalities.

It would be very exciting to see some other species evolve high-level intelligence. However, there are some clear obstacles to it happening naturally. First: as we have just discussed, humans would almost certainly need to be extinct. Second: yes, evolution is amazing, blah-blah-blah, but it really does take its sweet time to get anywhere. We just haven't got the kind of time it would take.

So, we need a shortcut. Our best bet for improving the intelligence of other animals would appear to be simply to make their brains a bit more like human brains. And one way of doing that would be the creation of human 'chimeras'. That means placing or growing human tissue – genes or cells, whatever it takes – inside other animals. We're already planning to grow human organs in other animals for transplantation. We think it should be possible to grow a pig with a human heart or liver, for example. No one is planning (or admitting to planning) to grow a human brain inside another animal, though, because that could lead to a whole raft of ethical issues. However, we might be able to tinker with the chimp brain and shift it more towards the human end of the scale.

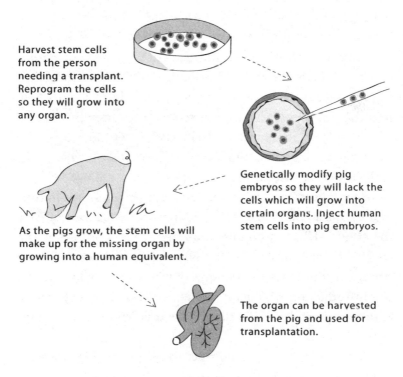

Harvest stem cells from the person needing a transplant. Reprogram the cells so they will grow into any organ.

Genetically modify pig embryos so they will lack the cells which will grow into certain organs. Inject human stem cells into pig embryos.

As the pigs grow, the stem cells will make up for the missing organ by growing into a human equivalent.

The organ can be harvested from the pig and used for transplantation.

**How to grow a pig with human organs.**

The most obvious difference between the brain of a chimp and the brain of a human is the number of interconnected neurons – seven billion vs eighty-six billion. So a Route One approach to trying to increase the chimp's intelligence would be to somehow add a whole load more fully wired-in neurons to its brain. That could be achieved by tweaking the genome in order for the brain to develop more neurons during growth.

Or, as with the 2011 prequel *Rise of the Planet of the Apes*, you could introduce neural stem cells, the cells that grow into brain cells. The chimp brain will then recruit and use them.

In *Rise of the Planet of the Apes*, James Franco is working on a genetic treatment for Alzheimer's. This is a disease that kills off certain neurons in the brain, severely affecting memory and other cognitive functions. One way that it's hoped we might be able to tackle, and maybe even cure, it is by growing new neurons in the lab and then introducing them to the diseased brain. This 'cell replacement therapy', like all cutting-edge technology, has to be tested. We can't carry the early testing out on humans because, well, it's a bit risky.

That's why, in the film, Franco is using chimps. As our closest relatives, their brains are a decent cipher for the human brain and therefore likely to yield robust results. Leaving aside the ethical debate around testing such things on primates – which is illegal in the UK, but allowed on everything except chimps in the US – the big question is around what exactly the introduction of human genetic material will do to an animal, especially one that is already quite similar to humans in the first place.

That big question is getting some pretty big answers. Remember that *FOXP2* gene – the one linked to the development of speech and language in humans? Researchers have spliced it into the developing brains of some mouse embryos. And what do you know: these mice have grown up and, in certain conditions, have improved learning capabilities. They aren't suddenly talking, but their squeaks are slightly different from those of regular mice.

So that's a single human gene having a clear, testable effect on the cognitive ability of a mouse. And that's the tip of the iceberg. It may sound utterly fantastical, but mice have been engineered with half-human brains. Not just for fun, but again to study various human brain diseases. These transgenic mice have good old-fashioned mouse neurons – the cells that are involved with actual thinking – but then almost all of their brain support cells, called glial cells, are human. Glial cells don't conduct electrical impulses themselves, but they provide insulation for the neurons that do. So it's essentially a mouse brain with support from human brain cells.

These human glial cells are much larger than the mouse equivalent, and can coordinate neural signals much better. That can be seen in the results of tests on the mice. The researcher Steve Goldman says they were 'statistically and significantly smarter than control mice'.

So what's to stop this kind of approach being taken with monkeys? Or with our closest primate relatives? Well, the scientists themselves, it turns out. They've slammed on the brakes. Their fear is that if you put a large number of human brain cells into a primate brain, that might result in a creature that has distinctively human capacities. Researchers are quick to point out that the human-enhanced mice are not somehow becoming more human. It's simply that the human cells are 'improving the efficiency of the mouse's own neural networks'. But in primates it might be a different story. Even more so if you were to implant human DNA into primate embryos. The result might be an ape that is self-aware to the same level as we are. An ape that can suffer like we can. And we're rightly uneasy about experimentation on such a beast.

This is not just an abstract concern. We don't know of anyone who is currently working to enhance cognitive function in apes, but the technical capability exists. And experiments involving the introduction of human brain defects to monkeys are certainly continuing.

## Anyone for brain soup?

Neurons are the functional processing units of the brain. The more neurons, the greater the cognitive processing power. So you might reasonably assume that the bigger the brain, the smarter the animal.

Broadly speaking, that's about right. But it's also about the number of neurons. For a long time it was an accepted and frequently quoted fact that the human brain has 100 billion neurons. But when the neuroscientist Suzana Herculano-Houzel looked for the source of this information, she found nothing. It's almost as if it was just made up. Could it be that no one had ever actually bothered to count? It seemed so. Lazy scientists.

Undeterred, Herculano-Houzel came up with a method. She took a sample of grey matter and used an acid to dissolve away the neuron cell membranes, thus leaving a suspension of neural nuclei swimming around in a fluid. Mmmm: delicious brain soup. She then shook the sample to give a homogenous distribution of nuclei, and then counted the nuclei in a given volume. Some quick maths led her to find that, actually, the average human brain has eighty-six billion neurons. We're not as smart as we once thought.

Still, the human brain has an awful lot of neurons for its size. That's because there's something special about primate neurons: as primate brain size increases, the neurons stay the same size. This means that growing the brain increases its processing power. It's not true for other creatures. In rodents, for instance, neurons grow bigger with a bigger brain. So even if a rat had a human-sized, 1.3-kilogram brain, it still wouldn't be as smart. It's been calculated that in fact it would need a 36-kilogram brain to have the same number of neurons as a human. That's not possible, because it would collapse under its own weight. It would also need to be contained within a terrifyingly big rat: around eighty-nine tonnes, roughly the weight of a young blue whale.

The first ever genetically modified monkey was born in October 2002. He was a rhesus monkey called Andi.* Andi had a simple marker-gene inserted into his unfertilized egg. But this proved that it would be possible to insert genes associated with specific medical conditions. And it is. In 2008 a research team spliced the gene for deadly Huntington's disease into the DNA of macaque eggs. To be sure that the insertion had been successful, they also spliced in a marker: a jellyfish gene that makes a fluorescent green protein. Sure enough, five bright-green macaque babies emerged, although only two survived beyond a month.

In Japan – where opposition to primate research is minimal – the first monkeys with Parkinson's disease have already been engineered. The marmosets have had a single gene, linked to the disease, mutated in their genome. And they are exhibiting the telltale symptoms of the disease, including the characteristic tremors.

The scientists' justification is obvious: the incorporation of human genetic information into 'animal models' has the potential to find solutions to life-threatening human conditions. But, given that we are introducing human genetic

---

* It's not just a cute name: it's 'Inserted-DNA', backwards.

conditions to monkeys, and growing human cells in other animals, it may only be a matter of time before we start putting human brain cells into our closest relatives. Someone, somewhere, might soon make a real-life Caesar, the super-intelligent chimp in *Rise of the Planet of the Apes*.

We should offer a cautionary tale here that, although it involves chickens and quails, might give us further qualms about monkeying around with apes. Evan Balaban at Harvard took brain cells from quail embryos and injected them into the brains of egg-bound embryonic chicks. These chicks subsequently hatched and, to the untrained eye, looked pretty chicken-y. However, Balaban coated their beaks with luminous paint, not because he's a weirdo, but so that he could observe their head movements. And that's where it got strange. They were bobbing their heads like quails. And they were crowing like quails. Somewhere, somehow, those quail brain cells were taking hold.

And so the waters get murkier. How many human brain cells can be introduced to an animal brain before it starts exhibiting human characteristics or behaviours? Could a chimp with enough human cells develop human conscious-ness? And would we be comfortable with that? The answers to those questions are: 1) we don't know; 2) presumably; and 3) probably not. In other words, let's be careful, eh?

Well, that's cleared that up then. So, basically, we're only on top because we accidentally grew big brains, and the way we're using those big brains could lead us to lose it all to a charismatic, super-smart chimp.

Would that be so terrible? Caesar seemed to know what he was doing. More so than many of our current human leaders, I reckon.

 You'd happily elect a human-chimp chimera as your leader?

At least when the shit hit the fan, you'd know who threw it.

# 5
# Back to the Future

CAN WE TRAVEL IN TIME?

HOW DO WE BUILD A TIME MACHINE?

COULD YOU ERASE YOURSELF FROM HISTORY?

Which one of the trilogy is your favourite?

 The first one, obviously.

But which do you mean by first? Chronologically, *Back to the Future III* is first. It's set in the Wild West, in 1885.

Well, obviously, you insufferable pedant, I mean the original. The Enchantment Under the Sea Dance, the Libyans, the loser peeping-Tom dad in the tree – it's a classic.

Ah, the weird dad. Not as weird as the actor playing him. Crispin Glover went on to make semi-pornographic art films and do a world tour of his PowerPoint presentation 'Crispin Hellion Glover's Big Slide Show'.

What the hell is that?

According to his website, 'A one hour dramatic narration of eight different profusely illustrated books he has made over the years.'

He lives alone now, doesn't he?

Yup.

This is surely a film that needs no introduction. The 1985 time-travel masterpiece has been around so long that the 'distant future' date of its sequel, 2015, has already been and gone. Michael J. Fox's role as Marty McFly, a teenager who gets transported to 1955 and accidentally undermines his parents' romance – threatening his very existence – has sparked myriad inconclusive late-night conversations about the paradoxes inherent to time travel. Less dissected is the 'flux capacitor', the time-travel mechanism fitted into a DeLorean car to turn it into a time machine. All it needs to function, says the inventor Doc Emmett Brown (played so memorably by Christopher Lloyd), is 'One point twenty-one gigawatts' of energy. And, we'd suggest, a plausible mechanism that doesn't stretch the known laws of physics to breaking point.

*Back to the Future* is hardly innovative, though. Time travel is a staple of science fiction, and has been ever since Herbert George Wells created the character of the Time Traveller in his 1895 book *The Time Machine*. But it raises an obvious starter question, at least: **Can we travel in time?**

## Take me back

 Which character do you identify with most in *Back to the Future*?

Doc Brown, obviously. Wildly intelligent and misunderstood.

 But seriously, which character do you identify with?

I've just told you – the Doc.

 I'll ask one last time. Be honest.

Fine. I identify with Biff.

 That wasn't so hard, was it?

Time is a weird thing. We know this because of Einstein's* special and general theories of relativity. As we saw when we looked into *Interstellar*, both of these show that time can be distorted, slowed and accelerated.

As Doc Brown would surely be able to tell you, the special theory came first. It was part of Einstein's 'miracle year' of 1905, when he published a series of astonishingly profound papers that changed the way we view many of the most familiar concepts in physics.

The central point in special relativity is that the speed of light is constant. What does that mean? Simply that when a car is moving past you with its headlights on, the light travels at the speed of light, $c$, relative to you. Not $c$ plus the speed of the car. Equally, if the car is reversing away from you, the light from the headlights still reaches you at $c$, not $c$ minus the car's speed.

If a constant speed of light doesn't sound that radical, that's because you haven't worked out the implications. Speed is distance covered per unit of time. So keeping $c$ constant actually means messing with distance and/or time. In Einstein's universe, measured lengths and intervals of time must change depending on movement. It's insane.

---

* The scientist, not the dog.

The distance thing is relatively dull. It simply means that if Rick flew past Michael, Superman-style, at close to the speed of light, Michael would measure Rick's height (well, length) as significantly less than Rick's own measurement of his grand extent. Rick would only have to travel at around 40 per cent of the speed of light for Michael to measure Rick's height as being equal to his own. Rick would dispute the result, of course, because his own measurement of his own height will always be the same majestic six foot five.

OK, so it's a *bit* weird. But an apparent ruler malfunction is nothing compared to the weirdness of comparing your measurement of elapsed time with a measurement taken on board a spacecraft that's moving past Earth.

In order for the physics of the universe to be the same for all observers, time passes more slowly on board that space-craft than on Earth. And that's true however you choose to measure it. Say Rick stayed on Earth, and Michael boarded the spacecraft and blasted off into outer space at close to the speed of light. The onboard clock runs more slowly than Rick's watch back on Earth. But it goes deeper than that. Michael's biology also runs more slowly than Rick's. He would literally age more slowly. Michael wouldn't notice it – it would all feel entirely normal – but if he returned to Earth after just under eighteen months of travelling at 99 per cent of the speed of light, he would have closed the ten-year age gap between them.

If we're being sticklers for the science, it doesn't work quite like that in practice, because of acceleration and deceleration and turning around, but you get the idea. This notion of the linear, unchangeable passing of time is old hat.

So, what can we do with Einstein's gift? Well, where special relativity is concerned, not a great deal. All we can do is try to

## Your mum...

Time travel makes things complicated – sometimes too complicated. It wasn't easy getting *Back to the Future* a green light, for example. Although luminaries such as Steven Spielberg were fans of the Bob Gale/Robert Zemeckis script, apparently there were plenty of reasons to hold off. Disney's reason was pretty special: the studio thought the plot was too risqué, what with Marty's mother trying to nail him in every other scene.

*Back to the Future* is not the most genetically risky time-travel film in history, though. 2014's *Predestination* (which stars *Gattaca*'s Ethan Hawke) surely has to take that prize. The central character uses time travel to creates lots of different versions of himself. And it's not straightforward: he is his own mother, father, son and daughter... This gender-twisting plot feels cool and modern, but it was actually based on a 1954 Robert Heinlein short story called 'All You Zombies'. Back then, some people felt it was just too much: the story made an editor at *Playboy* feel so queasy that he turned down the chance to publish it.

travel really, really fast. That takes us into the future of those who aren't travelling with us, like the time-travelling Michael making himself the same age as Rick.

The only people who have really done this are astronauts. Living in orbit for extended periods of time involves circling the Earth at speeds phenomenally greater than the speed we are moving at when on the surface of the Earth. Six months aboard the International Space Station, for example, gains you 0.007 seconds relative to your friends back on Earth. If you could sit on the GPS satellites, which orbit at around 14,000 km/h, you'd gain several microseconds every day. So far, though, the best anyone has done is gain 0.02 seconds. That's the best party chat of Russian cosmonaut Sergei Krikalev, who was in orbit for 803 days. Let's be honest: it's probably not enough to keep him in drinks at the bar, is it?

General relativity offers a better chance of time travel. For a start, time passes more slowly in a more intense gravitational field. On Earth, that means you age more quickly, the further you are from the centre of the Earth. Living on top of a skyscraper, for example, ages you faster. In fact, simply being tall is enough to accelerate decrepitude. As they go about their lives, Michael is going to gain about fifty billionths of a second on Rick, if they both live to eighty.

To be fair, that's probably nothing that Rick's skincare regime can't handle, and it's hardly going to take anyone back to the future. For that, you need what physicists call a 'closed time-like curve'.

Einstein's general theory of relativity says that the universe plays out its story on a stage, composed of space and time, that we call spacetime. We learned in *Interstellar* that this stage is decidedly bendy. Any mass and energy distorts spacetime, and if the mass and energy are concentrated enough, those distortions can become pretty extreme.

It's these distortions in space that cause planets to travel along the curved trajectories that we call orbits. A little harder to digest is the notion that time can bend, too, causing things to travel in odd temporal fashion. But the truth is, if you bend time far enough, you can create a loop where you keep coming back to the same moment in time. This is a closed time-like curve.

The first person to work out the maths of this was the Austrian mathematician Kurt Gödel. He presented it to Einstein in 1949, in a review of relativity's impact. It is safe to say Einstein wasn't terribly impressed; his take on it was that it was unlikely ever actually to be possible, given all the physical constraints on the universe.

In a way, Einstein was right to be sceptical. Gödel had done his calculations in a universe that was rotating and not expanding, and as far as we know our universe *isn't* rotating and *is* expanding, meaning that you couldn't get Gödel's naturally occurring closed time-like curves in our universe.

However, the idea of travelling through time remains possible. In theory, you can create closed time-like curves for yourself, and you don't necessarily need a flux capacitor to do so. General relativity tells us that all we need to do is bend spacetime radically enough to create a loop in time. Once you have that, you walk the loop and revisit the same moment in history as often as you want. As any old Doc, from Brown to Brooks* to Einstein, would tell you, creating such a loop requires a highly concentrated dose of mass or energy. And there are several ways we could create one of those. So far, so good, huh? Yes, we can travel in time.

It's probably worth pausing for a moment at this point, and offering a small caveat on all this positivity. We are about to enter the '*WHAT? You need WHAT?*' part of this module on the physics of time travel. As in '*WHAT? You need a neutron star?*' or '*WHAT? You need a source of hypothetical negative energy?*' or '*WHAT? You need a wormhole through spacetime*

* HOWZEEEEE!

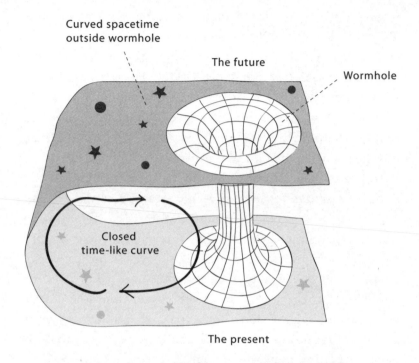

Curved spacetime
outside wormhole

The future

Wormhole

Closed
time-like curve

The present

**How to use a wormhole to travel through time.**

*that's anchored to the aforementioned neutron star?'* We appreci-
ate that these are not, generally, 'in stock' items and usually
have to be ordered in specially. However, it's not absolutely
impossible. That's what we're saying.

Right, with your expectations suitably addressed, let's
proceed with the next question: **How do we build a time
machine?**

# Blast from the past

 Favourite means of time travel?

The TARDIS. I always love how it's bigger on the inside.

 Boring – too obvious. For me, it's the unexplained mechanism in *12 Monkeys*, where it sometimes goes wrong.

I do like that quote: 'Science ain't an exact science with these clowns.'

 I'm fond of the film's idea that the early experiments sent people too far back and they were hailed as prophets.

Yes, you'd love that, wouldn't you?

 I honestly believe that I don't get hailed enough in the present day.

One of the biggest questions ever asked about time travel is this: if it's possible, where are all the visitors from the future? It's a good question, and one of the reasons that 400 people turned up to the Massachusetts Institute of Technology at 10 p.m. on Saturday 7 May 2005. It was billed as the 'Time Travellers' Convention' and was meant to be a gathering for visitors arriving from the future.

The rationale is surprisingly simple. If you hold the meeting and ensure there is a record of it, some people in the future who had access to a time machine would find the record. What could be more fun than a gathering of all such people at one temporal and spatial location? The advert for the event asked time travellers to bring proof of their futurity: 'Things like a cure for AIDS or cancer, a solution for global poverty or a cold-fusion reactor would be particularly convincing as well as greatly appreciated.' In a lovely nod to *Back to the Future*, the organizers arranged for a DeLorean to be sitting there onsite, in case people in the future had seen and appreciated the movie, or maybe even been inspired by it to go on and build a time machine.

So, how might they have done it? We have had a few ideas over the years. In the spirit of the movie, let's explore these chronologically.

The first suggestion for time-machine engineering is a simple one: a very long cylinder. That doesn't sound too difficult, does it? Unfortunately, the length of the very long cylinder stipulated by this machine's inventor was very long indeed. Oh, you want the precise length? Well, since you asked: infinite. Yes, that does sound a *little* difficult.

In 1976, American physicist Frank Tipler did some maths with Einstein's equations and worked out that a really, really heavy and infinitely long cylinder rotating really, really fast would warp space and time enough to create a closed timelike curve. Do we need to add that no one has ever really tried to turn this project into reality?

Next up was the brainchild of a man who glories in the name

## How to build a time machine

It's notoriously difficult to describe how time machines work, which is perhaps why we tend to get so little information onscreen about how to build one. As we've said, Doc Brown's flux capacitor requires 1.21 gigawatts of power to make time travel possible. We know a little more about Dr Who's TARDIS (Time and Relative Dimension In Space): it's a Type 40 time and space machine from the planet Gallifrey, powered by a combination of black-hole singularity, mercury, the rare ore Zeiton 7, a trachoid time crystal and artron energy. H. G. Wells's time machine is a bit more steampunk: its creator is an expert in 'physical optics' who creates 'a glittering metallic framework, scarcely larger than a small clock'. There's ivory in it, and 'some transparent crystalline substance'. And a quartz rod. And two white-handled levers. It's not a lot to go on, but it's way more than we know about how the phone-booth time machine works in *Bill and Ted's Excellent Adventure*. Finally, there's Hermione Granger's Time-Turner, which features in *Harry Potter and the Prisoner of Azkaban*. Here, finally, we have a full explanation. It uses magic – specifically an Hour-Reversal Charm.

J. Richard Gott the Third. Gott is also a physicist, and his idea was far more practical. And still impossible. Gott's idea involves hypothetical material known as cosmic string. It's a super-dense strand of matter, less than an atomic nucleus-width in diameter, that some cosmologists think might exist somewhere in the universe. If it exists, it would have formed as a result of the dramatic and traumatic processes that gave birth to the universe in a Big Bang.

Cosmic strings would provide a kind of natural time machine. They are extremely high-density defects in space that, placed side by side and moved rapidly apart, would create a time loop. All you have to do then is travel round the loop to revisit the same moment in history. Needless to say, no one has ever got their hands remotely near any cosmic strings, so this method is also not looking *entirely* promising. It's time to bring on the wormhole.

This beauty is an invention of Kip Thorne (yes, he of *Interstellar*). He invented it as a means of time travel for Carl Sagan's science-fiction novel *Contact*, where intelligent aliens send humans a message. The wormhole – essentially a means of quickly travelling across vast expanses of the universe – provided a way of visiting the aliens' intergalactic projects to learn more about them: the ultimate school trip.

Thorne's idea works like this. You find a natural tunnel through spacetime called an Einstein–Rosen bridge. Hold back on your scepticism: these *might* actually exist. The idea, first suggested by Albert and his pal Nathan Rosen in 1935, is that there could be connections between the cores of two black holes. After all, space and time break down at the core, which is better known as a 'singularity'. So what's to say all the singularities aren't connected, each a portal to a different region of spacetime?

If you can isolate just two black holes with a mutual connection, you have a 'wormhole' that connects two regions of space – or time. Obviously, finding one that starts where you are, and ends where you want to go to, is another challenge, but we're talking details now, really. One idea is to anchor it (somehow) to a neutron star, which has a super-intense gravitational field that slows time down. Get the far opening of the wormhole close enough, and you can allow a time difference to evolve between the two ends. When it's big enough, you can step into one end and emerge from the other end in a completely different time.

Actually it's not quite as simple as that, even. Space is kind of like elastic, and will resist being stretched, especially to breaking point. So if you stretch it enough to create a wormhole, you have to fight to keep said wormhole open. For that, you need something physicists call 'negative energy', and they're not sure it actually exists in our universe.

But apart from that, it's all quite straightforward, isn't it? It's maybe worth mentioning at this point that not every time-travel solution involves outer space or ridiculous practices such as spinning an infinitely long cylinder, or playing double Dutch with hypothetical cosmic strings. Ronald Mallett's idea is to use lasers. Where every other attempt at time travel warps space with mass on a cosmic scale, Mallett has worked out how it might be done in a laboratory on Earth. He even thinks we'll have a working time machine this century.

Mallett's determination to conquer time travel began on the death of his father from a heart attack. It was preventable, with lifestyle changes: if only he could go back in time, the ten-year-old Ronald reasoned, he could have warned his father. That's when he started reading H. G. Wells's *The Time Machine* and working harder in physics class. Now he's a professor of general relativity at the University of Connecticut. That's what motivation can achieve.

Mallett is so keen to make time travel a reality that he never wanted to mess around with ideas that can't be translated into reality. Now he has worked out how to create a ring of light so intense that its energy warps the space and time around it into a circle. In other words, there are closed time-like curves within the laser's path. Though you couldn't put a person into his time machine in its current design, he says it could be used to send messages encoded on subatomic particles back into the past. Mallett, who is in his seventies, is now fund-raising to build the machine. Hopefully he'll get some money soon, because there is serious Hollywood interest in turning his story into a blockbuster movie.

We forgot to mention what happened at the MIT time travellers' party. Did we need to? Of course no one actually turned up with proof they were from the future. But then, as Tina Fey pointed out ahead of the date, on *Saturday Night Live*, people from the future would have known if the party actually sucked. That means it must have been rubbish: that's why they didn't go. But hold on, this goes round in circles for ever, doesn't it? It would only have been rubbish because people decided not to go – because it was rubbish. Because, actually, no one went... WHAT? It's this kind of brain-befuddling reasoning that prompts our third question. Time travelling raises all sorts of difficult quandaries and confusing paradoxes. That's why Michael J. Fox ends up almost disappearing from his family photo – all record of his existence being slowly wiped out by his meddling in the past. Is this right? **Could travelling through time erase you from history?**

## The trouble with time

 OK, imagine you've got a time machine: where are you going?

BACK TO THE FUTURE

7

To kill my grandfather. Isn't that what you're
supposed to do, so you can prove it can't be done?

 Don't you want to go and see the future? After all,
you're not going to be around much longer.

Actually, I think I'll go and kill your grandfather.

 But then this book wouldn't exist.

I'm not so sure about that.

In one of the movie's most iconic scenes, Marty plays the Chuck Berry song 'Johnny B. Goode' at the Enchantment Under the Sea Dance. Marty would be able to tell anyone who asked that Chuck Berry wrote that song. But in the movie, Chuck first hears it via his cousin Marvin, who holds up the phone while Marty is playing. So who wrote the song? It's time for us to contemplate some of the most fascinating, mind-melting paradoxes in the whole of physics.

'I myself believe that there will one day be time travel because when we find that something isn't forbidden by the over-arching laws of physics we usually eventually find a technological way of doing it.' That's not just any old nut-job talking. That's David Deutsch, one of the biggest brains on the planet. He's a quantum physicist and the man who drew the first blueprint for a quantum computer. If he believes time travel is possible, because the laws of physics don't forbid it, that's a good reason to back the flux capacitor on Kickstarter.

However.

'It seems that there is a Chronology Protection Agency which prevents the appearance of closed time like curves and so makes the universe safe for historians.' Stephen Hawking wrote this in an academic paper in 1992. Informally, he calls them the 'Time Cops'.* These are the various things, such as the need for negative energy to keep a wormhole mouth open, that stop us actually doing any time travel into the past where we might alter the established facts of history.

Who to believe? Deutsch or Hawking? There is one classic time-travel scenario that always gets rolled out to try and answer this question: the grandfather paradox. It's pretty simple. Imagine that you travel back into the past, track down your grandfather before he got jiggy with your grandmother (sorry if that makes you uncomfortable, but it happened). Once you find him, you kill him. Now he can't conceive your mother or father, and so you can't exist to kill him.

*Back to the Future* contains a beautifully crafted version of this paradox. Marty's arrival in the past has caused his mother Lorraine, still a schoolgirl, to develop a massive crush on him. Now he has to make sure that, against the new odds, his hapless dad-to-be George gets together with Lorraine at the school dance. The more unlikely it looks, the more Marty and his siblings fade from the photograph that he has conveniently brought back from the future. It's the grandfather paradox, but done with panache. And, of

* Nothing to do with the 1994 Jean-Claude Van Damme film, our favourite review of which said, 'For once, Van Damme's accent is easier to understand than the plot.'

course, some questionable chemistry, what with that fading and reappearing image. But let's not quibble on that front.

- Show Stephen Hawking the end of the universe ✓
- Visit dinosaurs ✓
- Kill grandfather ✓
- Revisit 2017 to check on self ✓
- Visit 1889, find and kill Hitler as a baby ✓
- Visit 1945, check consequences of killing Hitler, stop screaming 'What have I done? What have I done?' ✓
- Revisit 1889, set up permanent guard to stop time travellers killing Hitler ✓
- Visit December 1980, buy shares in Apple ✓
- Annually - prevent the making of Back to the Future IV ✓

**A time traveller's to-do list.**

Stephen Hawking must have watched the movie with his heart definitely not in his mouth. Stephen doesn't believe there's any way you could change history. You didn't kill your grandfather, because you exist. If you go back to kill your grandfather, you'll find that one of any number of things happens to prevent the act. Your time machine will malfunction. The gun will jam. You'll slip as you take your one shot. The man you kill turns out not to be your genetic ancestor after all... You get the picture. This is 'chronology protection': whatever you attempt, the universe causes you to fail.

The easiest method of chronology protection, of course, is that you can't manage to travel back in time to start with. The idea that the laws of physics mean you can't make a working time machine is Hawking's go-to get-out for the grandfather paradox. That's why, he says, all the means by which we might achieve it involve seemingly impossible physics. However, there is one branch of physics where the seemingly impossible – things being in two places at once, for example – happens all the time: quantum physics. So maybe what we need is a quantum time machine?

All that swinging infinite cylinders around, or entering neutron star-anchored wormholes, is about creating a loop in time in the universe described by general relativity. But there is a more fundamental theory than relativity: quantum mechanics. We met this before, in our look at *Interstellar*. It describes the very smallest of worlds, the reality inhabited by photons of light and the subatomic particles that make up the substance of our being. Here, the rules are very different, and Hawking's Time Cops would be operating way beyond their jurisdiction.

What Hawking's complaint all boils down to in the end is that time seems to flow in one direction. That means a cause must always precede an effect. You can't have something happen without someone or something making it happen *before it happens*.

## saurs are off the menu

pected roadblock. According to the laws
an't go back any further in history than the
n the first time machine was created. The reason
ere are no rips in spacetime earlier than the first
ore there is no entry point. So you can't go back
t a dinosaur, or mate with a Neanderthal, or warn the
s passengers that it might not be quite as unsinkable
e captain says. Though this is disappointing, there is
upside: it shows that Stephen Hawking isn't right about
verything. He once pronounced that a good reason to think
time travel must be impossible is because we haven't yet met
any time travellers. But until someone invents a time machine,
there are no entry points. So an absence of visitors from the
future doesn't mean anything.

There's even more good news. With the kinds of physics
experiments we're starting to perform, we may be approaching the point in history when time travel becomes possible
and those hordes start arriving. In 2008, two Russian mathematicians pointed out that the Large Hadron Collider at the
European Centre for Nuclear Research (CERN) in Geneva could
become the world's first time machine. That's because it would
have the power to create mini black holes. Because the
smashes together tiny particles at huge speeds, the highly
concentrated energy involved can bend space and time
enough to create a time portal through which time travellers
could pour. As yet, we haven't spotted any mini black holes in
Switzerland, but it remains a possibility. If you're in Geneva
and you see a vast crowd of people looking like they've arrived

But quantum physi...
a quantum pheno...
mockery of cause...
ence was first...
too 'spook...
proved him...
suggest that we...
space and time.

Here's a lightnin...
two photons of light. You...
means. One is to create them i...
gent' crystal, where the atomic...
into two photons that have closely...
properties. Another would be to kno...
a very precisely controlled way. Afterward...
entangled.

What does that mean? It is a weird but absolutely...
feature of quantum physics that two entangled particles...
some of their properties. The classic one is 'spin': you c...
think of one having clockwise spin prior to entanglement,
and the other having anticlockwise spin. After entanglement, they both have some of each other's spin properties.
You can't find out exactly what that means, because the spin
is actually undefined until someone measures it. But once
you measure the spin of one, the result biases the result of a
subsequent measurement of the other's spin.

So far, so good. But here's where it gets really weird. After
the first measurement, the second particle's spin is immediately predictable. Even if the two are so far apart that
information from the first measurement would have to travel
faster than light – impossible in every law of physics – to reach
the second one and 'set' its spin.

from the future (and it's not just a coachload of Japanese tourists), do say hello. That said, be careful. They may have come back to murder their grandparents – and you might be one of the intended victims.

Physicists who do these experiments are repeatedly flummoxed by them. There is some mechanism behind entanglement – they talk about a 'correlation' between the results of the measurements – that defies all our understanding of how information passes through time and space. It's not just about spatial separation, either; there are experiments that show entanglements can be created and manipulated through time rather than space.

This is all completely mind-blowing, so what are we to make of it? One of the most eminent people in this field, Nicolas Gisin, once put it like this: 'There is no story in space and time that tells us how the correlations happen. There must exist some reality outside of space-time.'

The existence of a reality outside of spacetime implies that this universe is not all there is – and therefore that quantum weirdness does give us a loophole for time travel.

If you ask David Deutsch about that loophole, he'll say it's something to do with the Many Worlds interpretation. The idea here is that quantum particles such as photons can be in two places at the same time because their existence is not singular. There are worlds in which the photon is *here*, and there are worlds in which it is *there*. Some of our weirder quantum experiments, such as the famous double-slit experiment where a quantum particle simultaneously passes through two widely separated apertures, show us a kind of 'interference' between these worlds.

Or that's how Deutsch sees it. There are many different interpretations of these kinds of experiments, and we're not going to go into them all here. But the Many Worlds interpretation, which is gaining in popularity at the moment, says that we have an existence that moves between these worlds. That means there is no philosophical problem with time travel as such. We are not in danger of jeopardizing our existence by travelling to an earlier time, because we are actually travelling to another universe as well as another time.

If Michael uses a quantum time machine to go back and kill his grandfather, according to Deutsch he will enter a branch of the 'multiverse' where he wasn't actually conceived or born. So, whoever he kills, it isn't the guy who conceived one of his parents. At that time, in that branch, he has no forebears.

None of this is worked out in any detail – there is no truly coherent quantum theory of time travel. We realize it's not the most satisfying answer of all time. In fact, quantum time travel probably raises more questions than it answers. When your conscious self slips into a new world amongst the many, for example, what happens to the copy of you in another universe? No one knows. But maybe photographic records of it start to fade… and maybe the flux capacitor is some kind of quantum laser technology that ties together Ronald Mallett's work with David Deutsch's ideas. Maybe someone eventually worked out how to do it, then travelled back in time and gave the information to… No, hang on, that's not gonna work…

Did you know H. G. Wells married his cousin? So if he used the time machine to go back and kill his grandfather, he could get a two-for-one on paradoxes and would kill his wife's grandfather at the same time.

 That would definitely have wiped their kids off the family photo. If genetics hadn't done it already.

There weren't any kids. Herbert George divorced his wife to marry one of his students, and then went on a bit of a spree. He once said, 'sex is as necessary as fresh air' and 'every bit of sexual impulse in me has expressed itself'.

 Not an image I want in my head. Moving on, what have we learned? Time travel is possible; building a time machine is...

Difficult?

 Difficult-ish... Unless your local hardware shop has an infinitely long cylinder in a cupboard out back.

'Oh yes, here it is, just next to the paradox solution...'

And we could indeed erase ourselves from history, but only in a parallel universe where it didn't matter.

I think we've just worked out how to revive your TV career.

# 6
# 28 Days Later

SHOULD WE BE SCARED OF VIRUSES?

HOW CAN WE PROTECT OURSELVES AGAINST INFECTION?

CAN A VIRUS TURN YOU INTO A ZOMBIE?

It's so weird seeing London completely deserted. It must have been a nightmare to film.

It was. Danny Boyle had to employ an army of pretty girls to charm motorists into getting off the roads. And even though the police shut the M1 for two hours, the crew only got one minute's useable film.

That was the least plausible thing about *28 Days Later*, actually. Empty motorways? Never.

But everybody's dead or zombified.

Yeah, but there'd still be roadworks. It's a fundamental law of existence. Zombie apocalypse or no zombie apocalypse, there will be roadworks on the M1.

Maybe you love zombie movies, maybe you hate them. Either way, this one's different. It's (kind of) plausible and it's certainly iconic, with its deserted London streets and horrific red-eyed monsters. But it's also not exactly flattering to science.

The zombie state is induced by an engineered virus called Rage. Scientists have been infecting chimps with the virus, then making them watch TV footage of violent events as a precursor to developing an antidote to aggressive urges. Quite understandably, animal-rights activists are unhappy about this, and a trio of activists break into the facility to free the chimps (you'd think security would be tighter). Unfortunately, one of the activists gets bitten, and zombified. And away we go…

*28 Days Later* is far from the only movie where a virus is the real monster. There's *Contagion*, *12 Monkeys*, *I Am Legend*, *World War Z*, *Outbreak* and many more… We clearly have a fear of viral pandemic, which is one of the phenomena Danny Boyle wanted to explore in the film. So let's begin with a straightforward question: Is that a healthy fear? **Should we be scared of viruses?**

## The insider

Cillian Murphy drives me mad at the start. He's so ponderous, so slow to grasp the depth of the shit he's in.

It's almost like he's never seen a zombie movie.

Having seen most of them, I still don't know what I'd do to survive.

I can't help thinking it might just be better to get bitten. You know, get it over with.

And enjoy the company of the zombie horde?

Exactly. It'd be the most friends I'd ever had.

We had better start with a description of what a virus is, and what it does. First, a virus is a biological organism that... WHOAH! We've already gone outside the bounds of science. No one can agree whether a virus is biological or chemical. In other words, we don't know whether it is alive or not. That might seem ridiculous, but there is a loose set of criteria for defining life, and viruses don't fit all of them. Yes, they reproduce themselves, but they can't do it without the help of another organism. In other words, they aren't autonomous beings, able to float around their environment and get on with being a virus all alone. Being a virus is all about freeloading on other organisms. Maybe, way back in evolutionary history, viruses were free-living organisms that somehow lost their ability to live solo. But these days, they need things like us.

OK, let's start again. The basic components of a virus are fairly simple. First, there's a bit of DNA or its sister molecule RNA (you'll meet these again when we look at *Gattaca*). These molecules contain the instructions for building a copy of the virus. In order to create that replica, though, the virus needs the biochemical machinery inside a biological cell. Getting into the cell isn't easy, because biological organisms have defence mechanisms, such as enzymes that will chew up foreign DNA. So the viral instructions are enclosed in what's called a capsid shell: a protective layer of innocuous proteins. In some viruses, this is supplemented by an envelope of material that has been stolen from the ancestral DNA of the host. This all helps it slip by unnoticed.

And that's pretty much it. A virus can afford to be such a minimalist because it only has one job: to reproduce. The havoc it causes is not intentional. In fact, it's a bad virus that destroys its host: why do that, when the host can keep you alive and give you more resources?

The idea that viruses aren't always cold-hearted, deliberate, focused killers is backed up by the discovery that around 8 per cent of human genetic material is actually virus genetic

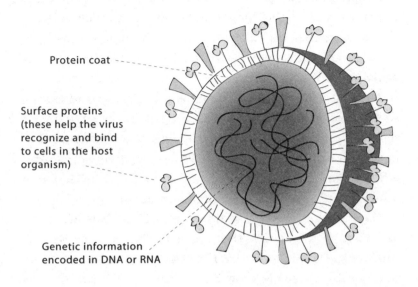

Protein coat

Surface proteins (these help the virus recognize and bind to cells in the host organism)

Genetic information encoded in DNA or RNA

**A typical virus.**

material. Clearly our ancestors became infected, and the viruses inserted some of their genes into ours. When we reproduced, it reproduced them too, though not in a form that gave us viral infections. Researchers suspect that, if they were to carefully analyse all our DNA, we might even find ourselves to be half virus.

That's not a bad thing: it's useful stuff. For a start, the changes it produced in other creatures throughout evolution's history almost certainly helped them to adapt to life in new niches. Viruses are also vectors for 'horizontal gene transfer', where organisms evolve through swapping bits of DNA: viruses are therefore part of the story of life. In humans, we know that viral DNA in the genome of a

developing foetus provides protection against certain infections carried in the mother's bloodstream. Genes from
ancestral viruses also exert control over some of our stem
cells, directing the genetic switches to produce certain types
of tissue.

A recent discovery shows that this cohabitation and mutual
support between viruses and their hosts has been going on
for a very long time. The pathogen known as Mimivirus
contains seven genes that also exist in every living organism.
They are part of the 'universal core genome' – a collection of
around sixty genes found in all life on Earth.

Having made the case that viruses are just 'one of us', we
also have to admit that they can be a bit of a problem at
times. After all, if you've ever had a cold (and we're betting
that you have), you'll know how annoying a viral infection
can be. And if you've ever had Ebola (we're hoping that you
haven't), you'll know how terrifying and, unfortunately, fatal
viral infections can be. So, if the virus needs us to give it a
full and rich existence, what's up with that?

Quite simply, it's the focused, singular drive towards replication that's to blame. First, the virus enters the body, via
the airways (welcome, influenza!), carried in the saliva of
an insect bite (come on in, yellow fever!) or by landing on a
cut or scrape on the skin or on mucous membranes (hello,
herpes!). Once inside, the virus attaches itself to the surface
of a cell and makes its way across the cell membrane. It does
this in various ways, depending on the virus. One is by simple
burrowing – as happens with the polio virus. The HIV virus
fuses with the cell membrane and gets pushed through to
the interior. Influenza profits from the cell's reaction: the cell
engulfs the virus, giving the attacker easy access to all its
machinery.

## The world's biggest viruses

The top three giant viruses all have more than one million base pairs in their genome. The appropriately named Megavirus, discovered in a sample of water taken from near the Chilean coast, is the biggest. Its genetic material has 1,120 protein-coding genes. Don't worry about this giant, though: it is thought only to infect marine bacteria.

That said, they thought something similar about Mimivirus, the third-largest virus. It has 979 protein-coding genes, and was discovered in the water from a hospital cooling tower in Bradford. Initially scientists said it only infected amoebae, but when it was being investigated in a French research laboratory, it managed to infect a technician, who went down with pneumonia. After this event the French researchers agreed to upgrade the research facilities housing Mimivirus to 'Biosafety Level 2'.

In the middle of these two giants sits Mamavirus, which has 1,023 protein-coding genes in its genome. It too infects amoebae, but has a peculiarity: it has its own parasite. When Mamavirus sets up its virus factory inside an amoeba, the tiny viral parasite – dubbed 'Sputnik' by its discoverers – invades this factory and uses its machinery to replicate itself. Rather pleasingly, Sputnik disrupts Mamavirus's operations to the point where the bigger virus produces deformed copies of itself. In other words, Sputnik makes Mamavirus unwell.

Now the fun begins. Once inside, the virus releases its genetic material, which hijacks the replication mechanisms and gets itself reproduced many times over. The host cell usually dies, because its own routine has been so rudely disrupted, and the newly created virus particles break out to infect new cells, beginning the process all over again. Evolution has led the most successful viruses to create conditions that also get them dispersed to infect another organism. So the rhinovirus that gives you the common cold also makes you sneeze, sending 20,000 virus-laden snot droplets out into the air other potential hosts are breathing. Ebola's evolved method is a much blunter instrument: it is, essentially, to liquefy you so that its newly made virus particles can easily escape from the confines of the human body and spread into the wider world.

It's this need to keep reproducing by finding new hosts that's your main problem as a host. With a common cold, the evolved methods are annoying for human beings, but rarely fatal (any longer). With Ebola, rabies, smallpox and a host of others, they bring disaster.

But human beings are also part of the problem when it comes to viruses. We could argue the pros and cons for days, but many people – OK, scientists and terrorists – see viruses as potentially valuable tools. That has led to the situation where our biology laboratories are full of engineered versions of the little buggers.

It can't be denied that they are useful. We can use viruses to deliver genetically engineered solutions to medical problems, for instance. Understanding viruses also helps us understand our own biology and improve our resistance to infection. There are attempts to use viral DNA's control of genetic switches to fight cancer – which makes sense, now that we understand that some viruses (human papillomavirus, for example) actually cause cancer. We also have to study viruses if we want to know how to create vaccines – something we'll get onto in a moment.

Unfortunately there's also the issue of weaponized viruses. As we've seen, the thing about successful viruses is that they spread themselves around extremely well. Some are naturally good at this, and others are not. Whether that can be changed is an extremely important question.

Right at the beginning of *28 Days Later*, a scientist who's been working with the infected chimps makes a terrible stab at justifying the laboratory's work. 'In order to cure, you must first understand,' whines David Schneider's character – somewhat unconvincingly, it has to be said. To give Schneider the benefit of the doubt, maybe it was meant to be unconvincing. Maybe the scientist had never had to justify his research before.

When it comes to viruses, scientists do the funniest things. Take the scientific work on Asian bird flu H5N1 virus, for example. H5N1 is a nasty piece of work – you really don't want to catch it. Fortunately, though, it's quite difficult to catch. As with the Ebola virus, you have to be in close contact with a bird to stand a chance of picking up your own set of pathogens, and it doesn't usually spread from person to person. Bioterrorists would love to change that. That's why there are active programmes to find out how difficult it is to 'weaponize' H5N1 by making it able to spread, like the common cold, by airborne transmission – a process known as aerosolization. That way, governments can decide whether they need to protect their populations against an immediate threat and start work on a vaccine against an airborne strain.

The only way to find out whether terrorists might succeed in aerosolizing H5N1 is to try it. In other words, to do the terrorists' work for them.

As it happens, it can be done, and it has been done – by a group of Dutch scientists, back in 2012. It might surprise you to learn that these scientists kicked up an almighty stink when it was suggested they might not want to publish their techniques in the open scientific literature. It might surprise you even more to learn that they carried out their research in a lab that wasn't 'maximum security'. It was designated biosafety level 3+, one level below the maximum.

We know what you're thinking. *WHAT?* Strangely, scientists don't actually have a great track record at keeping their most dangerous work as safe as it can be. There are spills, for example. In 1978, researchers at Birmingham University accidentally released smallpox virus into their building's ventilation ducts. The virus found a host in journalist Janet Parker, who was working on the floor below. As a result, Parker has gone down in history as the last recorded victim of the disease. It's not the only such case. In 2004, two researchers working with the SARS virus in China somehow managed to infect themselves with it. They carried the infection out of the lab, and seven other people became infected. One of those unlucky seven – one of the researchers' mothers – died from the infection.

Concerned researchers did some calculations of the likely future of this kind of thing, and worked out that there was an 80 per cent chance that the Dutch aerosolized H5N1 virus would escape its lab within four years. With those odds, there's every reason to stay vigilant.

For all the concern, it's still unlikely that human negligence, stupidity or malevolence will cause an actual global viral pandemic. It's arguably more likely that a previously unencountered virus will make its way into the human environment at some point and cause havoc. If that sounds far-fetched, it isn't: there are a *lot* of viruses around. A teaspoon of seawater contains around one million virus particles. In fact sampling the oceans has revealed there are millions of viruses that have never found their way onto land and into our laboratories for identification. That means there is a host of possible new ways of getting ill, and maybe dying. Which makes it sensible to ask our second question: **How can we protect ourselves against infection?**

## Catch me if you can

The stupidest moment in the film must be when Cillian Murphy shaves his beard off. Why would he do that? Beards are great.

That's not why it's stupid. He was shaving dry, and lacerated his face. That made it easier for a splash of zombie blood to infect him with Rage.

So beards not only look good, they also protect you from infection?

And from being beaten up, as it happens. Studies show that people perceive men with beards as stronger and more intimidating.

We are in an evolutionary arms race. Pathogens want to use us for their own ends and are developing new ways to do that all the time; we are constantly fighting back with improved immune systems. That's why last year's winter flu jab is no good this season – you need a shot of something new. The influenza virus can't afford to stand still. It has to mutate to survive, to bamboozle your ever-alert immune system.

You should be proud of your immune system. It's a defence mechanism so complex that we struggle to understand its extraordinary efficacy. Developed over millennia, it uses a host of tricks to identify and neutralize threats to your body's well-being in a war where the collateral damage has to be kept to an absolute minimum. Usually there'll be a slight temperature hike and some lethargy – maybe a runny nose, some muscle aches or pain when swallowing. But that's a small price to pay, considering what's going on.

You actually have two immune systems. One is 'adaptive', and is composed of a variety of different cells that circulate in the bloodstream. These cells make antibodies and other molecules whose role is to recognize specific proteins – usually associated with bacteria, parasites and viruses – and latch onto them using a lock-and-key mechanism.

The adaptive system is the source of acquired immunity. When our bodies first play host to a pathogen, certain cells learn to make the antibodies that will shut it down. These cells respond to the distress signals from infected cells, and

## Home Alone 2: Lost in New York

In many ways, a disease that you carry for weeks, spreading the infection widely before anyone knows you've got it, is much more deadly than Rage. These days millions of people travel across the world every day. That's why viral diseases such as bird flu and Zika have become so dangerous: they have relatively long incubation times, allowing people to travel and communicate the disease unchecked, well before symptoms alert anyone to the danger.

Quarantine restrictions are often seen as the answer. Anyone who might have been infected is identified and isolated. It's a strategy that has been in use since ancient times, but it's not foolproof – especially in the modern era. Restricting travel is difficult. With the 2013 Ebola outbreak, stopping people from moving around West Africa was seen by some as an external imposition that adversely affected the region's economy. Recommendations that the 2016 Rio Olympics should be cancelled or moved because of the risk of Zika-virus infections were met with claims that such a response was simply unthinkable.

Sometimes, though, quarantine gets enforced. In 1972, the Yugoslav government used martial law, at the request of the World Health Organization, and quarantined a village where smallpox infections had been discovered. The restrictions worked: the villagers experienced the last smallpox infection in Europe.

In 28 Days Later the British Isles are thrown into quarantine. Naomie Harris's character, Selena, claims the quarantine has been unsuccessful and that Rage cases have been seen in Paris and New York. She's wrong – but not for long. In the sequel, *28 Weeks Later*, the infection arrives in Paris. Not because someone let a raging red-eyed zombie through, but because of something even scarier: a carrier who doesn't get ill. We have seen asymptomatic carriers of a number of viruses: HIV, typhoid, Epstein–Barr and chlamydia, for example. You have been warned.

release chemicals that can latch onto the virus (or bacterium or whatever) and halt its fun. The successful defensive cells then reproduce, creating successive generations of cells that become shoot-on-sight killing machines for that particular pathogen.

The 'innate' immune system is different. Though not able to tailor a response to a particular pathogen, it attacks anything foreign or unusual. Its cells, which include the white blood cells known as scavenger cells and T-cells, will detect and destroy bacteria. Others, known as natural killer cells, check tissue health by looking for changes in cell surfaces that indicate the presence of tumours or viruses. Then there are enzymes that chemically mark threats. Like the laser guides for missiles, they can bring in scavenger cells and other immune-system artillery. The enzymes also get busy by dissolving the cell walls of bacteria and digesting the outer layer of a virus so that it is more vulnerable and less well disguised. This is the part of the immune system that causes inflammation reactions, often associated with fever, which are a side-effect of the ongoing battle to heal you.

Great as our immune system is, it benefits from a helping hand in the form of vaccines. When you get a vaccination, a disabled, dead or severely hampered version of the pathogen is being introduced into your bloodstream. Your immune system finds it, and develops the antibodies required to kill that particular pathogen. Those antibodies stay with you, making you immune to that specific threat.

Vaccinations are one of the greatest success stories of human history. They now avert two to three million deaths every year. Measles vaccination, for instance, has saved more than seventeen million lives so far this century.

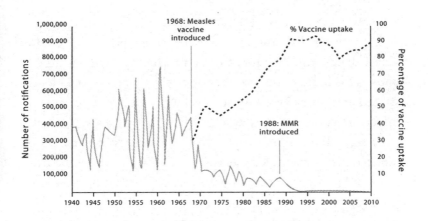

**Introducing the measles vaccine halted the stop–start of breakouts and virtually eradicated the disease in England and Wales.**

However, vaccines only work if the pathogen doesn't change too much. If its physical form evolves beyond a certain point, the antibodies don't recognize it. So if you're a dangerous zombie-creating virus, evolution is your friend.

This is the big problem with HIV: it evolves super fast. A single virus is able to spawn literally billions of copies in just twenty-four hours and, because it's not a great copier of itself, those copies will have subtle differences between them. Some of the differences give the virus an advantage in defeating the host's immune system. Even worse, two different versions of HIV can combine within a host cell, creating another new variant. This high natural variability made HIV extremely difficult to treat; it quickly became resistant to early drug-based treatments, and to anything the host's immune system could throw at it. Its rapid evolution, which varies depending on the host, has made vaccine development very difficult, too.

Fortunately, a new generation of anti-HIV drugs, known as antiretroviral therapies, has been enormously successful. These drugs stop the virus from reproducing itself within the body. This reduces the number of viral particles in the bloodstream and other fluids to the level where it is not possible to infect someone else. Though the virus isn't destroyed, it isn't winning, either. This is one of the greatest achievements of modern medicine; as long as you have access to these medications, HIV is no longer a death-sentence. Getting access to the drugs is an economic and social issue, though – and the same problem occurs with vaccine development, as the recent Ebola outbreak made clear.

We first saw Ebola in 1976. It is transmitted through fluid exchange – through sex, or open wounds, breast milk and any other direct contact where body fluids might flow from one person to another. It won't turn you into a zombie, and it won't make you attack others. But an infection will give you fever and malaise as your body tries to fight it, and the virus causes diarrhoea, vomiting and leaking body fluids in order to escape and infect a new host. The infamous, horrific bleeding from the eyes, though not common, is all part of Ebola's strategy to take over the world. It occurs because the virus breaks down the body's various mucous membranes, including those in the eyes, which hold blood vessels close to the body's surface. It also stops blood from clotting, so that once the oozing starts, it doesn't stop. Unless you're lucky enough to have some natural immunity (if you've caught Ebola, 'lucky' is probably not how you'd describe yourself), you're dead from multiple organ failure within a week.

Horrific as it sounds, Ebola was not initially considered enough of a threat to Western countries to merit a vaccine programme. In the 1970s, the US military looked at it and decided that, if they were working in an Ebola-prone zone, good hygiene and minimal physical contact with carriers would keep it under control. A vaccination development programme, with all the palaver and cost that involves, simply didn't seem worthwhile.

It is largely thanks to the benevolence of non-profit groups and charities such as the Wellcome Trust that we have a vaccine now (and, if we're being cynical, the threat of infected people getting on a plane and coming to Europe and America). Once we tried making a vaccine, it actually took next to no time. The international effort started in 2014. By the end of 2016, results were in on trials of a vaccine given to nearly 6,000 people in Guinea. It worked, and 300,000 doses were on order within months. The next time Ebola rears its ugly head, it will meet immediate and effective resistance.

What about *28 Days Later*'s Rage? It was based on Ebola, and produced many similar symptoms – the burning red eyes, for example. So how plausible are its other effects: **Could a virus turn us into rampaging zombies?**

# The fast and the furious

> I love the detail towards the end, when the zombified Private Clifton sees his reflection in the mirror. He looks a bit confused. Does that mean he's not self-aware?

> I think you might be reading too much into that scene.

> I don't see why. It's a valid question: are zombies self-aware?

> Are *you* self-aware? Listen to yourself. Everyone else is terrified for the child who's clinging on to the back of the mirror, and you're wondering whether the set-up constitutes a valid experimental protocol.

Ah, how ironic. The Rage virus in *28 Days Later* was originally designed to aid the discovery of a drug that can reduce violence. Unfortunately, when its DNA was combined with Ebola,* evolutionary mutations made it do exactly the opposite. Cue rioting, rape, murder, starvation – and a generally very un-British British Isles.

Ebola infection certainly doesn't make anyone rampage.

---

* We learn this is what happened in the sequel, *28 Weeks Later.*

As we've seen, it destroys the body, making victims weak and unable to do anything for themselves. This is our general reaction to illness, in fact: one of the first tasks of the immune system is to make sure we put all our resources into fighting an infection, rather than going to the office or getting on with rewiring someone's house. Biomolecules called cytokines knock you off your feet and take away your appetite, making sure that you don't waste energy, even on something as passive as digestion. So could the opposite happen?

Well, yes. We know this because there are infections that will change human behaviour. Some do it in subtle ways that are unlikely to create a ruckus. But others are a little bit frightening.

Let's start with the gentlest of changes, just to ease us in – the pathogen known as ATCV-1 chlorovirus, to be precise. Originally we saw this virus as something that infects algae. However, when it turned up in a study of the microbes found on and in the bodies of psychiatric patients, microbiologists at the Johns Hopkins University School of Medicine in Baltimore got curious. They carried out some tests and found that 43 per cent of sampled people were carrying it.

That's quite worrying, because ATCV-1 has brain-befuddling properties. It slows your brain's visual and cognitive processes by around 10 per cent. It reduces your attention span. Mice infected with the virus have lower attention spans than uninfected mice and can't find their way out of mazes anywhere near as quickly. They also lose curiosity; they just aren't that interested in exploring new objects. Presumably, if a more virulent form were to emerge, we would get even dumber – more zombie-like.

### Deadliest catch

Rage killed its millions, but other infections have done worse work...

Smallpox, caused by the viruses *Variola major* and *Variola minor*, is the only human infectious disease to have been eradicated from the natural world. It has been charged with the deaths of up to 500 million people in the twentieth century alone. No one knows how many it killed before that.

Bubonic plague gave us the Black Death, which wiped out one-third of Europe's population in the fourteenth century. Something like seventy-five million people died in the pandemic. The bacterium responsible is still with us, occasionally causing an outbreak.

Spanish flu gave us an appalling viral outbreak at the end of the First World War. Roughly one-third of all the people on Earth were infected, and 50–100 million people died from its effects.

The malaria parasite kills around two million people per year, most of them children under five years old. A vaccine is in development, but progress has been unconscionably slow because most of those affected – in Africa, Asia and South America – are not in a position to pay for treatment.

HIV has killed more than twenty-five million people. Despite our success with developing drugs that can control the effects of the virus, its attack on the immune system continues to create an enormous death toll in regions of the world where the drugs are not freely available.

Tuberculosis, a dangerous bacterial infection, kills between one and two million people every year. The bacterium is hosted by around one-third of humanity and causes around ten million people to fall sick annually. Influenza kills half a million people annually. Its rapid evolution means that flu vaccines, though effective, have to be redesigned every year.

Typhus, a bacterial infection, killed around three million people between 1918 and 1922 alone – soldiers have been particularly affected by this plague. The bacterium is carried by lice, and is particularly virulent in conditions where hygiene is restricted. Today, it is largely under control, killing only one in five million people worldwide.

OK, we'll live with that (and maybe we already are living with it). But we're also living with our cats, and they too carry parasites that, some researchers claim, can affect our behaviour. You probably already know that the *Toxoplasma gondii* parasite (a single-celled organism) in cat faeces is dangerous to pregnant women because it can be transmitted to the foetus. However, an infection can hit anyone. Estimates vary, but it's thought that around one-third of humanity is carrying a *Toxoplasma gondii* infection. Its reported effects are subtle: it can make us slow, angry and – somewhat weirdly – more sociable. If you're looking for something that, amped up, would create a lumbering but aggressive zombie horde, maybe you should look no further than your cat's litter tray.

The research on *Toxoplasma gondii* has thrown up some interesting findings. We know, for instance, that infections are more common amongst people with certain psychiatric disorders, such as schizophrenia and bipolar disorder. But even more disconcerting is the link to a condition called 'intermittent explosive disorder'.

Those with IED are prone to moments of uncontrolled aggression; one of the symptoms is an increased tendency to display road rage, for example. According to research by Emil Coccaro at the University of Chicago, these people are more than twice as likely as 'normal' people to be carrying *Toxoplasma gondii*. It's hard to be sure about how a cat parasite would cause rage, but one theory is that the infection triggers brain chemicals that overstimulate the brain's reaction to a perceived threat, or simply inhibits the processing pathways that rationally assess threats in the environment. And why? Maybe it makes the cat's prey unable to distinguish real threats from false ones, making the hunting process easier. This would be an example of symbiosis: a parasite and its host operating for mutual benefit. The prey is more sociable and interactive, easing the parasite's spread. It is also a bit confused and slow, making it an easy target for the host.

In humans, experiments suggest that *Toxoplasma gondii*'s effects are a little more complicated. Infections tend to make women more outgoing and trusting, while men become more introverted and wary of others. Everyone, though, had slower reaction times. If that is the route to a zombie apocalypse, then the uninfected should at least be able to hold their own in a fight.

We can probably do better. To create a zombie, you need radical behaviour change, something like what a parasitic fungus called *Ophiocordyceps* achieves. Found in the Brazilian rainforest, the fungus infects ants, releasing a chemical cocktail that turns them into tiny automata that have no control over their behaviour. After two days the zombie ants do the will of the fungus: they climb to a particular height where the temperature and humidity are optimal for fungal growth, and latch onto vegetation with their mandibles. Once they are fixed, the fungus kills them by releasing a chemical weapon, and then grows a spore-releasing stalk called a stroma out of the back of their heads, so that it can spread itself further. It's pretty gruesome.

1. An unsuspecting ant picks up a spore from the rainforest floor. The spore makes an enzyme which breaks through the ant's exoskeleton and gets inside.

2. Two days later the ant leaves its colony and climbs to where fungus growth conditions are best. It latches onto a leaf there, where it dies.

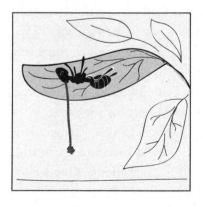

3. The fungus then grows a 'stroma' from inside the ant's head. This carries new spores, which fall to the ground to infect more ants.

**How to zombify an unsuspecting ant.**

As yet, no mind-controlling fungus has been found to infect humans, although TV commissioners are holding off from a third season of Rick's ITV2 show *Safeword*.

It's time to get to the one you've all been waiting for. If we're trying to weigh up the chances of creating Rage, we can't ignore the elephant in the room. Especially if it's infected with rabies.

Rabies is a truly terrifying disease. It is Ebola-like in the excruciating process of multiple organ failure leading to a slow, agonizing death. Its fatality rate in humans is almost 100 per cent. Around the world, it kills seventy-five children every day. But it doesn't, like Ebola, make people lie down and die quietly. It makes them rage.

Humans unlucky enough to be infected by the rabies virus can become wildly aggressive. They experience delusions and hallucinations, sweat and salivate wildly, and display an uncontrollable urge to bite others. It's a classic viral move: the virus accumulates in the salivary glands, and so biting is its best means of spreading to another host. Another set of symptoms – fear of water or liquids, plus uncontrollable muscle spasms when the victim attempts to swallow – add to the virulence. The best way out for all that virus-loaded saliva is through the open mouth, not down the throat. And if you can't swallow water, or even bear the sight of the stuff, you can't dilute the saliva's viral load. This virus really does know how to get the job done.

The rabies virus can control behaviour in this way because it makes its way into the central nervous system and the brain, where it causes swelling that affects behaviour, mood and motor functions. If you ever wanted proof that your free will can easily be bypassed, that a zombie-like state is possible, it's right there in the scientific literature on rabies. Like we said, terrifying.

So, what can we conclude about Rage? Even just a cursory glance through the symptoms of a few known infections gives us almost every component of the Rage virus. All we need to do is to put them together. So it's not hard to imagine a dystopian experiment that puts a fungus, some cat shit, saliva from a rabid dog and an algal virus together in a nightmare-inducing Petri dish and waits for evolution to do its thing. After a while you might have something interesting. Something that makes people sociable enough to leave their houses and get together in hordes, but weirdly aggressive, slightly dumb and slow, and as nippy as an excited terrier. They would also have the unenviable experience of being totally unable to control their own actions. Could we create Rage? It's not entirely impossible.

So we're right to be scared of viruses – especially ones that liquefy you. I'm comforted that we're getting really good at limiting the threat, though.

But what if something new evolves that can turn us all into zombies?

I'm still amazed that might even be possible. Appalled, actually.

Well, evolution generally progresses by accident. It would take a long time for that to come together.

Unless some accident-prone scientist gives evolution a helping hand.

Oh, come on, that's too unlikely.

 Really? Have you learned nothing from all this?

# 7
# The Matrix

ARE WE LIVING IN A SIMULATION?

CAN WE EXPERIENCE 'BULLET TIME'?

WILL WE EVER HAVE INSTANT LEARNING?

Fair play to the Wachowskis. The visual effects in this film are still extraordinary, nearly twenty years later.

And yet they couldn't make it look like Keanu Reeves can act.

Nice. Jealousy disguised as critique. Well, I would love to watch this film with you cast as Neo. In fact, I'd love to watch any film with you in it. Then we'd see who can't act.

Are you suggesting I'm jealous of Keanu Reeves? That's ridiculous.

Of course it is. I momentarily forgot that you too are a multimillionaire lady-magnet and Hollywood success story...

The year is 1999 – or so we think – and Keanu Reeves is living a double life. By day he's boring old computer programmer Thomas Anderson, but by night he's a hacker operating under the pseudonym Neo. Neo feels as if he is waiting for something, but he's not sure what. Could be a sign from God; could be an Ocado delivery...

The truth is, he's waiting for his DESTINY because he is THE ONE.* Unfortunately, that particular destiny comes with an eye-popping revelation: everything in the 'real' world is in fact a simulation.

Here's the scoop. At some point in the past, we've had a war with machines. The machines won and enslaved the human race (if you think that's implausible, skip straight on to the *Ex Machina* chapter; then you'll be straight back here). Cleverly, the machines have plugged all the humans into a

---

* If Neo were as good at anagrams as he is at hacking, he would have seen that ONE coming.

sophisticated computer program: the Matrix. This creates a virtual reality in which we are relatively content and don't ask too many questions about the unfortunate fact that we are living out our days immersed in vats of liquid while the machines use us as energy sources.

When Neo learns that humanity is just one big, deluded battery pack, he's not a happy (Duracell) bunny. Something has to be done... and it turns out that he's the man for the job. It's a great set-up, and our first question is obvious. Are the Wachowskis telling it as it is? **Could we be living in a simulation?**

## Déjà vu all over again

 It's Plato's Cave, isn't it?

What?

 Plato told a story about people chained up in a dark cave. Their world comprises the shadows on the wall in front of them, which are created by giggling, maniacal puppet masters outside the cave. Because the cave guys don't see anything but two-dimensional shadows, they think that's reality.

Could they not just look at the person next to them?

 What? No, their heads are fixed looking forward.

> And they don't remember having their heads clamped by someone three-dimensional?

 It happened a long time ago. When they were little.

> Oh, and these magic clamps don't require any adjustment or maintenance over the years? Anyway, they'd still be able to see someone out of the corner of their eyes. Or they could just talk to each other. Presumably someone sneezed or coughed at some point?

 This is why I'd rather live in the Matrix than share a world with you.

In the movie, Morpheus – Laurence Fishburne's character – gives Neo a profound insight. 'Real,' he says, 'is simply electrical signals interpreted by your brain.' But he's not the first to raise the point. The question of whether reality is real is an age-old concern.

Back in the seventeenth century René Descartes had a good long think about whether it was possible that he was just a floating brain, being systematically deceived. In his book *Meditations on First Philosophy*, Descartes imagined that an evil demon was feeding lies about the external world into his unwitting brain. Substitute 'sophisticated computer program run by machine overlords' for 'evil demon' and you've pretty much got Neo's predicament.

But even René was beaten to the punch. In the fourth century BC the Chinese thinker Zhuang Zhou had a vivid dream about being a butterfly. When he woke up, he couldn't help but wonder: was he now a butterfly dreaming that he's a man? Common sense suggests not, but he couldn't be absolutely certain. The argument that a butterfly doesn't have the cognitive power to dream up the human world doesn't really hold, because if you can't be certain whether you're dreaming or not, you can't reasonably say much about how smart a 'real' butterfly is.

More recently, various twentieth-century philosophers, including Gilbert Harman and Hilary Putnam, have considered the unpalatable notion of a Brain In a Vat (BIV). This is a thought experiment in which you imagine that you are a brain hooked up to a computer that can perfectly simulate experiences of the outside world (or at least, an outside world). If you cannot be sure that you are not a brain in a vat, then you cannot rule out the possibility that all of your beliefs about the external world are false. Gutted.

Which means that the most salient question is: Where are those electrical signals coming from? Are they being generated by an evil demon? A supercomputer? Or are they what we usually assume: responses to real-world stimuli?

The problem is, your brain is a sort of isolated thinkbox, sitting (hopefully) in your skull, in total darkness and silence. Everything your brain 'knows' about the outside world comes in, and out, via electrical signals running along bundles of nerves. The brain takes all of this sensory information and stitches it together to tell itself a story about the world – for example, you have a cup of tea in your hand and it's hot. In *The Matrix*, whatever signals would have been coming in from your hands or eyes are simply replicated by the computer.

And you still think you have a hot cup of tea in your hand, you fool. A scant consolation in the film is that your brain is, at least, still embedded in your pale, withered body, and not just swimming around in a jar.

So far, so unnerving. And then along comes Nick Bostrom. In 2003, four years after *The Matrix* was released, this Oxford University philosopher formalized an altogether more out-there idea: what if *nothing* was real and *everything* was simulated – *including your brain*?

Now there is no vat and no brain, just a simulation of consciousness. That might sound far-fetched, but stick with us, because old Nick has really thought it through.

First, he points out that our technological capabilities are improving at a helluva rate. The processing power of our computers is still increasing fast, and there's no reason to suppose that will stop. Second, we seem quite interested in running simulations: be they virtual-reality video games, or The Sims, or evolutionary models. Third, we are making significant advances in, and ploughing money into, the mapping of the human brain.

Bostrom concludes from all this that it is likely we will one day have the capacity to create incredibly detailed simulations that are, thanks to our knowledge of the brain, populated by beings who display all the signs of consciousness. This is slightly contentious – no one's quite sure whether we will actually be able to model consciousness – but let's say that we do reach that point. We might then use these simulations to 'rerun' history, and see how things might have been different, for example. It's something that historians and evolutionary biologists do all the time in thought-experiments. Question: How great would it be to do it for 'real'? To establish once and for all how life began... when consciousness arose... the origins of language... Answer: Really, really great.

If Nick is right, this all leads to three future scenarios. These comprise his 'simulation argument'. In the first one, the human race becomes extinct before reaching the necessary level of technological advancement. In that scenario, the simulations never occur. The second scenario is where we reach the required level of technical competence, but decide not to run these simulations, possibly because we become bored or decide it's unethical. Lastly, there's the scenario where we become those fearsome, tech-savvy über-geeks and go ahead and run the ancestor simulations (or whatever).

Let's look at each option in turn. The first is fairly bleak. Given that we don't feel terribly far away from implementing rudimentary simulations now, that option seems to suggest our end is nigh. The second option doesn't seem that likely. Humans are curious and, if we have the ability to do something, we tend to get on and do it. Also, if historical re-enactment societies are anything to go by, humans really enjoy raking over the past.

So that leaves the third option, Bostrom's simulation hypothesis itself, where humans – or possibly post-humans – are extremely interested in ancestral simulations (as we already are) and able to create ones containing conscious beings.

Because computing power would be so vast at this point, it would be possible to run loads and loads of these simulations in a very short timeframe. That would mean that you have the base-level reality humans or post-humans – the ones who created the very first simulations – and all the simulated humans within the simulations. The base-level reality consciousnesses will be far outnumbered by the simulated consciousnesses, which means that statistically, when considering our own consciousness, we have to concede that we are more likely to be simulations. Dammit.

Start here...

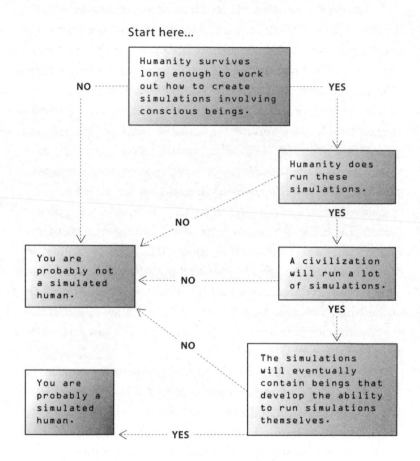

**Are we living in a simulation? You decide.**

So how could we prove any of this? One thing that might help is if we could find any 'glitches in the matrix'. The code isn't perfect, presumably: in the movie, for instance, a cat walks past Neo twice, a telltale sign that he's experiencing the simulation, not reality. An experience of déjà vu is a giveaway. An experience of déjà vu is a giveaway.

Alternatively, maybe the simulation will betray itself by being a little too coarse-grained at times: maybe we'll see gaps as well as glitches. These gaps might appear if the operators of the simulation are worried about having enough computing power, in which case they could just create a simulation that extends to what we can observe from where we are within the universe. In other words, a lot of the distant stuff could be very poorly rendered, like the painted backgrounds in early films. We'd never know. Equally, there might not be any need to render all the small stuff until someone looks at it – quick, they've got the electron microscope out! Program some particles to behave like particles! It's classic 'if a tree falls in the forest but nobody hears it, does it make a sound?' stuff. And a possible answer, if we're living in a slightly half-baked simulation, is No. It doesn't. Because the mob running our simulation is trying to save on computing power. So maybe, just maybe, we could catch them out.

The third way our simulation might be rumbled is much more upsetting, or it would be, if we knew anything about it: the whole thing might be powered down. The possibility of getting switched off, or of computing resources being reallocated, disrupting our normal function, is a valid concern. After all, the operators might get bored of us, or decide to run another, better simulation instead. And here's where it all gets quite depressing.

Assuming a huge stack of simulations within simulations, with the single base-level 'reality' right at the bottom, the density of simulations towards the top of the pile is much greater. You can think of it like an inverted pyramid, with each simulation holding up many others, each of which has more simulations piled on top. With every level making its own set of simulations, it's absolutely heaving with fake realities up

## Digital heaven

The futurist Robin Hanson published a book in 2016 called *The Age of Em*. In it, he discusses the possibility that before we fully crack artificial intelligence and are able to run vast simulations of the type suggested by Nick Bostrom, we might be able to copy ourselves digitally. He calls these digital copies 'Ems' and imagines individuals having an army of them, all off doing a whole host of jobs. It would certainly make multitasking easier. And assuming your Ems are in gainful employment, then you can just sit back, relax and enjoy the future.

If we are in fact living in a nested simulation, there are ramifications for the concept of afterlives. Humans have long been fascinated by the idea of resurrection post-death, and this becomes all the more pertinent if we are purely simulated consciousnesses, tracts of code. That would mean, as Hanson posits, that we could easily be replicated on another computer. Or, in the case of death in a simulated universe, the people running it have the option to re-create you, as an 'Em', in their higher-level universe. Like getting promoted from the Championship to the Premier League. Except that the Premier League might not be the top league. And you might get re-created again and again, rising up through the simulated universes. Although it's unlikely that the simulators above us will re-create and promote just anyone. Maybe only the good people...

there. So again, statistically, we're more likely to be towards the top. And because the whole thing is precariously balanced, it presents a clear and present danger: not only could our simulation be switched off, but if any simulation directly below us in our simulation family tree gets deleted, we'd be sent tumbling into the void, too.

There is just one saving grace. As mentioned, creating a simulated consciousness might just not be possible. Christof Koch, a hugely respected neuroscientist, has been studying consciousness for decades. He is now president of the Allen Institute, whose aim is to build a full schematic of every neuron (brain cell) and synapse (the connection between two neurons) in the brain. He believes that a physical machine like this, which mimics the structure of the brain, *could* be conscious – in his words, 'it would feel like something to be this computer'. However, he does not believe that a digital simulation, a kind of software model of the brain, would be conscious. He argues that a simulated consciousness would feel nothing, in the same way that a physical person can't live inside a simulated house, and in the same way that the Met Office's computers create simulations that include rainclouds, but never actually get their circuits *wet*.

Indeed, Koch reckons the fact that *we* feel things proves we can't be part of a simulation. But many others, including Bostrom, simply disagree. So we don't really have an answer. Ultimately, though, if we and our reality are simulated, would it really matter? Is it any less real? Many physicists who have no stake in the simulation argument believe that all the attributes and processes of the physical universe can be reduced to information-processing anyway, so what's the difference? A tree is still a tree, even if it's made up of digital code, rather than pure biology. Everything is still the same.

Except, perhaps, our perception of time. And here *The Matrix* scores again. One of the puzzles of our physical universe is that time appears out of nowhere: it's in our heads, but not necessarily a fixed component of the universe (cast your mind back to *Back to the Future*). And here the Wachowskis have messed with our minds even more. Arguably their greatest brain-boggle is not the simulation idea, but something that has come to be known as 'bullet time'.

Neo's mastery of the Matrix is so amazing that he isn't constrained by the clock that runs within the simulation. So he can opt out, and slow things down. This is what enables him to dodge the bullets fired by the agents who are trying to kill him. So, if you want to dodge a bullet (and who doesn't?), all you need is for time in the external world to pass much slower than it does for you. Which forces our second question: **Can we have bullet time?** Please?

## Time to die for

Did you read that Malcolm Gladwell book where he says that it takes 10,000 hours to master a skill? It's a lot, isn't it? I don't know that I want anything that much. Do you think you'd ever put 10,000 hours into anything?

Yes – into trying to sell as many books as Malcolm Gladwell.

It's not going very well, is it?

To be fair, I've probably only done 7,000 hours so far.

Yes, but I'm not sure you've actually got 3,000 working hours left.

Here's a revelation: cinema is all lies. When you watch *The Matrix*, you're watching a sequence of still images that your brain interprets as movement. You knew that, of course. But have you ever thought about what it means? The fact that movies work is just one indication that our brains are fooling us. And nowhere are they fooling us more than in our perception of time.

Time is a jerry-built, rickety edifice created by our brains. The jelly in your skull assembles all of the available sensory information – visual and auditory cues, for example – and creates an impression that accounts for the duration and sequence of events. So whilst it seems that life unfolds in a continuous spool, your brain is just putting together lots of snapshots of the outside world, exactly as it does when watching a film like *The Matrix*. Because of this, time actually runs differently for different people, depending on how long it takes for signals to travel through their body.

It's not easy to assign a specific value to the rate at which the human brain samples its environment. However, if we want to experience bullet time, presumably all we have to do is radically increase the sampling rate and recalibrate how our 'subjective time' (how long we perceive something to last) compares to 'objective time' (what our watch tells us about the passing of time). If our brain knows – or thinks it knows – that it will have $x$ frames of visual information per second, but then it suddenly doubles its sampling rate to $2x$ frames per second, it will interpret this as a duration of double the

## Past masters

'It's really important to live in the moment.' It's the kind of thing you'd hear from a self-help guru. Pleasingly, it's not possible, because we all actually live in the past.

This is down to the way our brain processes the sensory information from our nerves. The data is coming through at different speeds from different places, and is being processed by different bits of the brain. The brain then has to do some nifty 'temporal binding', where it edits and stitches everything together to make one neat picture.

An unexpected consequence of this is that the brain has to wait around for the slowest bit of information to arrive before it can perform the final assembly. The delay is roughly one-tenth of a second, but the exact delay will depend on your size. Michael is not as freakishly tall as Rick, and if something touched their toes simultaneously, it would take longer for the sensory information to travel from Rick's toes to his brain. Michael's short limbs have finally given him an advantage: he is living marginally closer to the present.

What's more, waiting for all the information to arrive is only half the battle. Your brain assumes that when you interact with the world, the corresponding sight, touch and sound are simultaneous. When you click your fingers, the feeling of doing so, the image of it happening and the sound of the click certainly seem simultaneous. But the brain has done some work to achieve that, using its own expectations about the incoming signals to present you with a sensible picture.

length. In other words, time will have appeared to slow down. Subjective time will have been altered, whilst objective time remains the same. Bingo: bullet time.*

Is it possible? Well, maybe. Flies sample the world more rapidly than us, which means that, relative to us, they are living in a world where time moves more slowly because they will observe motion on much finer timescales. And that is why we believe flies are able to dodge a newspaper so easily: to them, the newspaper is moving at a very pedestrian pace. Flies are experiencing their very own bullet time, all the time. Rolled-up-newspaper-time, you might say.

And it's not as if you haven't experienced something like bullet time. Many of us tell tales of moments – usually of danger or high stress – when time has appeared to move more slowly. Why? Is it possible that our brain has turned up the sampling rate?

It's a question that neuroscientist David Eagleman tried to address in an extraordinary experiment. He persuaded a group of volunteers to take a ride on a theme park's 'suspended catch air device'. This is, essentially, a drop from a fifteen-storey platform. It's terrifying – which is exactly what Eagleman wanted.

He asked his volunteers to retrospectively report the duration of their fall. He also asked them to watch other volunteers fall and estimate how long that took. The volunteers all estimated their own falls to be around one-third longer. This is a time-dilation effect, which suggests that time had subjectively slowed down for the terrified free-fallers. So far, so good.

---

* Using bullet time to improve your bingo performance seems a little wasteful, to be honest.

However, each of the fallers was wearing an instrument that Eagleman had designed with his student Chess Stetson.* It's called a 'perceptual chronometer'. Essentially, it's a watch that flashes up random digital numbers at an adjustable rate. Specifically, the perceptual chronometer flashes up, say, a red 83 on a black background, followed by a black 83 on a red background – an exact inversion of the previous display.

When two images arrive within a small window of time, something of less than 100-millisecond duration, the brain's collation program integrates the images. So if that second image – a negative of the first – comes up very quickly, the brain sees a blank.

The volunteers wore the perceptual chronometer on their wrists. Eagleman then adjusted the flash rate, so that he could establish each volunteer's perception threshold: the upper limit at which the volunteer was still able to read the number. Then he turned it up a bit. If time was really slowing down for them during the fall, and their temporal resolution was higher – more 'frames per second' – then they should have been able to read the numbers flashing at this increased rate.

Here's where it all falls apart. None of the volunteers could read the numbers during their plummet. This suggests that the fallers were not experiencing higher temporal resolution at all. So why did they all report that the fall had lasted longer?

It might be because danger gives us a special type of false memory. Under stress, a part of the brain called the amygdala takes charge, and the memories are laid down in 'high definition'. When the brain recalls this memory, it looks at the high density of data and assumes it must have taken a while to

---

* Scientifically proven to be the most American name ever.

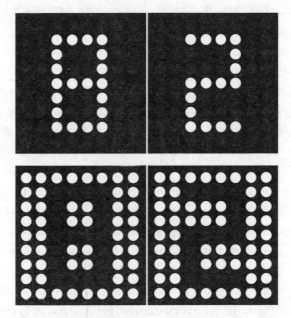

The perceptual chronometer flashes alternately between the numbers and their inverse.

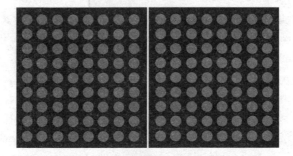

As this alternation becomes faster, the brain creates a 'null' combination of the two images and we can no longer read the numbers.

How the perceptual chronometer works.

record all that. You think – in Eagleman's words – 'Gee, that was taking a long time.'

If Eagleman is right, in moments of danger you're not like a fly. You can't avoid the danger, because time hasn't slowed down. You just have a heightened recall of the threat. It's like Neo remembering the bullets flying towards him in slow motion, but not being able to move: 'That bullet's going to hit me, that bullet's going to hit me. Oof! That bullet just hit me!'

When you think about it, it's the worst possible outcome: a strong, detailed memory of unavoidable disaster. But hang on a minute. None of this explains a very common recollection of brief, dangerous situations. We often express amazement at the sheer number of thoughts that passed through our heads and the actions we performed in what was, objectively, a split second. If, as Eagleman's free-fallers suggest, temporal resolution is not enhanced, and the slowing of time is just a trick of the memory, how come we react as if time was slowed for us?

Valtteri Arstila from the University of Turku in Finland has an idea that might save us. He has suggested that stress hormones, related to the fight-or-flight response, swiftly trigger a mechanism that massively accelerates the brain's processing ability and speed. This makes it feel as if the outside world has slowed down. Research done with people taking part in high-risk extreme sports suggests that some are able to 'turn on' this slow-time perception – to control their very own bullet time, in other words.

Even if this is true, the mechanism is not yet understood, so it's not clear how you're going to benefit – except by repeatedly skydiving off cliffs or taking up some other stupid, death-defying pastime. But there is hope for us mere mortals/sane people. In an experiment at Keele University, subjects were exposed to a ten-second burst of fast audio clicks (about five per second). Then they performed some basic mental tasks: arithmetic, word recall and target identification. They were 10–20 per cent faster after the clicks than before, suggesting that somehow the internal clock-rate of the brain had been accelerated.

We'll take that. It may not help us dodge bullets, but it might be nice to be able to shift through the mental gears on occasion. And that raises our third question. In *The Matrix*, Neo learns how to fight (and do myriad other things) through a plug-in interface that uploads skills modules to his brain. Could we ever do the same? **Will we ever have instant learning?**

## I know kung fu

Before shooting, all the lead actors in *The Matrix* had to read *Simulacra and Simulation* by Jean Baudrillard. Ever read it?

Of course I have. It's a 1981 classic that contains one of the most prescient views of our digital lives, predicting that people in affluent, successful societies will become more concerned about curating the public representation of their existence than they do about actually enjoying life.

Did it resonate with you?

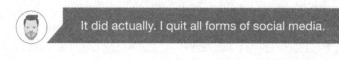

It did actually. I quit all forms of social media.

For how long?

Nearly five hours. But it felt like an eternity.

One of the most famous moments in *The Matrix* is when Keanu Reeves is plugged into a computer and has a martial-arts upload. After a few moments he opens his eyes and proclaims 'I know kung fu' in an unconvincing monotone. You might not want to be Neo, what with all those high expectations to do with being THE ONE, but you almost certainly wish you could learn stuff without making any effort.

If we want to short-circuit the brain's traditional way of learning things, we'd better understand exactly what it does when we absorb knowledge. Unfortunately, that's quite a tall order. When we learn, the physical structure of the brain is changed. The strength of connections between brain cells determines the quality of memory and the ease of recall, and so learning requires that neurons fire signals in patterns that create or strengthen particular synaptic connections between them. Gladwell's thousands of hours of practice slowly instils mental and muscle memory by forming these bonds between neurons. When learning something new, recall and memory will be strengthened by frequency – doing it loads of times – and recency – doing it consistently.

No surprise there: neuroscientists have seen this in mice. When two neurons in the tiny mouse brain interact regularly, they form a bond that allows more accurate transmission.

The opposite is true of neurons that don't talk much – their transmissions are often incomplete, which gives rise to patchy memories, or no memory at all.

The challenge, then, is to identify the patterns of firing that result in a specific moment of learning. And then you stimulate the brain to do that particular firing routine over and over, until the synapses are connected up just the way you need.

How might we do this? Well, at the moment the best hope is a technique called decoded neurofeedback (DNF). The basic theory runs like this: let's say that Rick can solve a Rubik's cube puzzle with ease, but Michael can't. Rick could teach Michael how to do it, but it'd take a while, and Michael doesn't like it when Rick is better at him than things. So, using an fMRI scanner, we measure Rick's neural activity as he solves the cube. We record this activity. Rick is now free to go and present a daytime quiz show or whatever.

Michael, who doesn't have a daytime quiz show to present, gets hooked up to the scanner. A computer algorithm analyses his neural activity and compares it to Rick's recorded activity. Now the algorithm teaches Michael to make his neural activity more and more like Rick's. It does this by presenting him with an image – say, a circle – on a screen in front of him. The disc will get bigger as Michael's neural patterns get closer to Rick's, and smaller as they get further away. The positive and negative feedback means that Michael's brain gets used to these patterns of firing and will be able to induce this perfect pattern of neural activity for the task. It's worth emphasizing that Michael need have absolutely no idea what he's learning to do. All he does is let his brain respond to the disc on the screen and, at the end of the process, he's a fully fledged Rubik master.

## How many ways to tie a tie?

Not many films can claim to have spawned the creation of several thousand new ways of tying a necktie. But the *Matrix* trilogy can.

In 1999 – coincidentally the year of the first film's release – a couple of mathematicians, Thomas Fink and Yong Mao, developed a notation for tie knots to demonstrate that there are only eighty-five possible knots. However, this conclusion came undone when a Swedish mathematician called Mikael Vejdemo-Johansson stumbled across a YouTube tutorial on how to tie a knot like the Merovingian, a character from the later *Matrix* films. He immediately noticed that Fink and Mao's work was missing this fancy knot. The fools!

So Vejdemo-Johansson did what anyone would do in that situation – he got around to rewriting the tie notation to accommodate the Merovingian style. He also changed a rule. Fink and Mao had decided that the maximum number of 'winding moves' allowed would be eight, because any more makes the tie embarrassingly short. But Vejdemo-Johansson figured that you could always make longer ties, and allowed up to eleven winding moves.

And so the maximum number of tie knots went up from 85 to more than 177,000. That's an embarrassment of riches; Fink and Mao can't show their faces in tie circles any more.

To be clear, DNF isn't yet advanced enough to teach something as complex as the process of solving a Rubik's cube. But it has been successfully demonstrated in the visual cortex, believed to be the simplest area to test. Professor Takeo Watanabe and his team at Brown University were able to use decoded fMRI to induce brain-activity patterns that matched a target state corresponding to a simple pattern of stripes. OK, so it's not as exciting as instantly learning to solve a Rubik's cube – let alone the whole kung-fu thing – but it's a start. Professor Watanabe's team were able to improve their subjects' visual performance. Even better, the improvement is long-lasting. So this rudimentary 'implicit learning' works and, according to Professor Watanabe, it could, theoretically, be extended to complex motor skills. Like, say, kung fu.

Impressive, huh? Of course it is. But there's a catch (of course there is). The brain-activity patterns associated with muscular movements are hugely complex. What's more, they will give rise to significant individual differences: Rick's neural code for solving the Rubik's cube is unlikely to be exactly the same as someone else's. In other words, our brains aren't identical in the way that computers are, and therefore a generic, standardized 'program' for any task may be unattainable.

All is not lost, though. Another technique, which involves stimulating the brain with a mild electrical current, appears to accelerate and improve learning. This can either be transcranial direct current stimulation (tDCS), which is a weak but constant current delivered to the brain via some electrodes on the skull (ouch!), or transcranial random noise stimulation (tRNS), which is a randomly fluctuating current. If you're squeamish about passing electricity through your brain, you should know that tRNS is apparently a bit more comfortable.

tDCS has been shown to improve people's ability to learn strings of numbers. tRNS, which is a bit newer, has helped improve a range of number skills: subjects using tRNS were able to memorize new equations and perform new calculations more quickly than the control group, whose electrodes weren't connected to anything useful. That's because tRNS seems to stimulate a part of the brain that we think has a role in mathematical cognition. Bizarrely, it also appeared to make the brain more efficient: the metabolic levels in the tRNS group were significantly lower than those in the control group.

These initial successes mean that more sophisticated cognitive training programmes will be developed in the near future. So how will we access them? In *The Matrix*, Neo gets plugged in. That seems acceptable, because we already have brain-machine interfaces working quite well. It might sound futuristic, but we can create electrical signals that we input into the brain, or transfer thoughts directly brain-to-brain.

In the late 1990s a Brazilian researcher called Miguel Nicolelis taught a monkey to control the position of a dot on a computer screen – first with a joystick, and then just with its thoughts. Let that sink in for a moment. A monkey... controlling a cursor... with its mind. Nicolelis has since used a similar interface to allow paralysed human patients to control prosthetic limbs. One of his patients used a robot exoskeleton to kick off the 2014 World Cup at the Corinthians Arena in São Paulo.

The next step involved brain-to-brain interfaces: feeding the signals from one set of firing neurons into another set, held in a different skull, and watching the outcome. Two plucky researchers sat in different rooms while wearing electroencephalography (let's go with EEG) caps that captured their brainwaves. The signals from one EEG cap were fed into the other, so when the first researcher imagined himself playing a video game and shooting, by *imagining* pressing the fire button, his brainwave got transmitted to the next room. The second researcher, in addition to the cap, was wearing a transcranial magnetic stimulation (TMS) coil, which emits

At first, the monkey learns how to move the cursor around on the screen by using the joystick. During this, the computer analyses the monkey's brain signals, matching them to the cursor movement.

Then, the joystick is removed. Now, the monkey can move the cursor just by thinking about it, as the computer is able to interpret the monkey's brain activity.

**How a monkey controlled a computer with its mind.**

focused electrical signals. It was positioned over the part of the brain that controls the movement of the finger. When the first researcher imagined pressing the fire button, his EEG cap would 'see' that and communicate with the second researcher's TMS coil. This would then emit a signal based on the EEG signal, and his finger would actually press the button. This is very unsettling for the person whose finger is being controlled, apparently, who cannot distinguish between commands being given by his own brain and commands coming from the external source. This is not like having a voice in your head: as the second researcher said, 'The first time I didn't even realize my hand had moved. I was just waiting for something to happen...'

Freaky. But clearly there is potential to input body movements into people via a brain interface. Learning can be stimulated. Movements can be stimulated. Maybe one day someone will plug themselves in and declare, a few moments later, that they know kung fu.

This is all making my head hurt. I should just have uploaded it.

And one day you will.

I can't wait. So, bullet time is hard to access, but if you're a fly, or an adrenaline junkie, or an adrenaline-junkie fly, it sort of exists already. Instant learning is coming to a private school near you soon. And we might well be living in a simulation – but good luck proving it.

Worth a try, though. How about some civil disobedience to make our post-human overlords break cover? How do you think they would react if we started trying to crash their computer systems?

 Don't... Can we stop talking about this, please?... IT WAS HIM, IT WAS HIM. I'M VERY HAPPY, THANK YOU. DELETE HIM.

You're no Neo, are you?

# 8
# Gattaca

ARE WE MORE THAN JUST OUR GENES?

HOW ACCURATE IS GENE-BASED PREDICTION?

SHOULD WE USE GENETICS TO CREATE THE PERFECT HUMAN?

I remember watching this film in the cinema and feeling extremely smug because I understood why, in the title credits, the letters A, C, G and T were highlighted.

Hard to imagine you being smug, Rick.

I felt even more smug when I realized that the name of the film is itself just a DNA sequence, made up of the four letters that represent the elements of DNA: adenine, cytosine, guanine and thymine.

I expect you were one of the very few people to spot that.

Really?

No.

In *Gattaca*, Ethan Hawke's character Vincent has a serious problem. He was conceived in a very old-school way – his mummy made love to his daddy, got pregnant... you know the kind of thing. In *Gattaca*'s world, this is not OK. You're supposed to use genetic screening and IVF to ensure that a baby is as perfect as possible.

When Vincent popped out, the disapproving doctor took a blood sample from his heel and, through near-instantaneous DNA analysis, reeled off a list of potential problems and genetic conditions this newborn would be dealing with in the years to come. It included a heart issue that gave him a life expectancy of 30.2 years. Sobering stuff, which presumably put a dampener on any baby shower that his parents had planned.

This was in stark contrast to Vincent's perfect younger brother Anton, with whom his parents took no chances. After consultation with their friendly local geneticist, they picked the best embryo for the IVF, the one representing the ultimate combination of their genes. As the geneticist says, 'The child is still you, just the best possible version of you.'

There are two castes in *Gattaca*: the genetically enhanced 'valids', and the supposedly genetically inferior 'in-valids'. And then there's a whole lot of 'genoism': discrimination against someone because of their genome. Our first question, then, has to be the one raised by the premise of the film: **Are we more than just our genome?**

## You've got male

 Wanna hear some *Gattaca* trivia?

Do I get a choice?

 Of course not. The pre-release publicity for the movie included full-page newspaper ads for 'Children Made To Order'. It had a list of traits that you, the diligent parent, could pick: gender, stature, skin colour, athletic prowess, intellect… And, as you'd expect, loads of absolute goons called up to place their order.

That's completely understandable. Have you met my kids? I'd definitely want a better outcome next time.

Fair point. But you have to remember the scientists would still only have your genes to work with. They can't perform miracles. Which reminds me: do you want to know what my favourite bit of the film was?

Was it the bit where the doctor says that he wishes his parents had ordered him a dick like Vincent's?

No... Well, yes.

The philosophy in *Gattaca*'s world is that genes maketh the man. That's why Vincent's genetically superior brother, Anton, is astonished to lose a swimming race against his sibling. 'How are you doing this, Vincent?' he says, exhausted and near to drowning. 'How have you done any of this?'

*Gattaca* was made in 1997, perhaps the peak of the 'our genes will unlock the mysteries of humanity and disease will be consigned to the history books' delirium. James Watson, of DNA-discovering fame, was saying ludicrous things like, 'We used to think our fate was in our stars. Now we know, in large measure, our fate is in our genes.'

It was a time when the Human Genome Project was in full, expensive swing, and promising all sorts of things. The director of the project, Francis Collins, was banging on about 'the first draft of the human book of life'. Strong stuff. As it turns out, too strong. The Human Genome Project may have produced a book, but it was a prohibitively long and complicated read, and written in a language we don't fully comprehend.

In chemical terms, a gene is a string of molecules. The human collection of genes – the human genome – is a long chain constructed from just four basic chemical building blocks: adenine, cytosine, guanine and thymine. Within the genome these four chemicals, known as A, C, G and T, are not just the letters in our movie title: you can think of them as letters that form a very long set of words that constitute the instructions for building a human being.

Our genome is essentially a set of instructions then. There are only four letters in its alphabet, but the whole thing is about three billion letters long.* Those letters are grouped into around 20,000 units that we call genes, each one of which encodes the instructions for making a protein, or a set of proteins.

The genetic letters bind to each other in specific ways: A on one strand pairs with T on the other, and C pairs with G. These pairings are held, like the rungs on a ladder, on long chains of what are essentially just sugar and phosphate molecules. The whole thing forms into a long double helix that we know as DNA.

A copy of the genome is held in the nucleus of almost every cell in the body. When it's time to make new biological tissue, an army of molecular machines uses these instructions as a basis for its work.

Is that all there is to building a copy of your biology? No – far from it. While genes are important, they can't be everything, for a number of reasons. One is that, as mentioned in our *Planet of the Apes* chapter, you share 98.5 per cent of your genes with chimps, and they're a different biological species.

---

* Don't get cocky about this 'large' number. An ugly brute called the marbled lungfish has 133 billion letters in its genome. Go figure.

That means the huge difference between building you and building a beast from another primate species is contained within 1.2 per cent of your genome.

Then there are all the differences between humans, which are encoded in just 0.1 per cent of our genomes. Two billion nine hundred and ninety-seven million of the three billion letters in Rick's genome are exactly the same as in Michael's. In case you're not genetically predisposed for maths (which isn't a thing, by the way, but we'll get to that kind of misunderstanding in a bit), what we're saying is that only three million base pairs are different between you and the person seated next to you.

Also, there's the 'junk DNA' to consider. Most of the long double helix of humans – a massive 98 per cent of it – doesn't code for building proteins at all. It's just random-looking arrays of A, C, G and T. Different species have different amounts of junk DNA. There's a growing suspicion that these 'useless' sequences do *something*, but no one knows quite what that something is.

What's more, the usefulness of your genome is not just about the genes. It's also about the molecular machines, such as enzymes and proteins, that do the work of building your cells.

We won't get into all the devilish details of molecular biology here, but consider the fact that our genome controls the production of some 100,000 proteins. These proteins – the production of which is different in different kinds of cells – are behind differentiation, which makes skin cells different from neurons, which are different from blood cells, and so on. But while our genes are controlling protein production, the cell is also controlling the activity of our genes, changing the way proteins are produced. It's another classic chicken-and-egg kind of situation. Another important factor is how our genes

Bladderwort: 97% genes; 3% junk          Humans: 2% genes; 98% junk

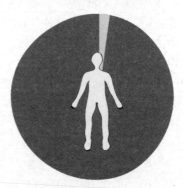

Black grape: 46% genes; 54% junk          Nematode worm: 29% genes; 71% junk

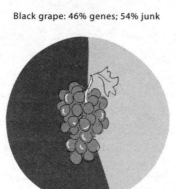
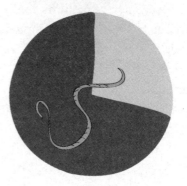

**Different species have different amounts of 'junk' DNA.**

interact with *each other* during development – the activity of one can affect the activity of another. And the same gene can do very different things within different genomes. Then there's 'epigenetics': the heritable effects of the environment on the activity of our genes.

## Epigenetics

The results of environmental factors on gene function are often called 'epigenetic'. By 'environment', we mean a variety of sources. Within the body, certain chemicals can attach themselves to the genes, inhibiting their normal function or activating them. Psychological factors such as stress can produce such 'epigenetic marker' chemicals. Then there's the external environment: pollution products such as smoke particles have been shown to have epigenetic effects linked to asthma and other allergies. Food also has an effect: methyl groups of carbon and hydrogen atoms can make their way from our diet to our genome, attaching themselves at sites where they turn genes on or off, changing the body's protein-production processes.

Epigenetic markers can have positive or negative effects on our health – and, possibly, on the health of our offspring. This is a relatively new area of biology, and there are still a lot of unknowns, but evidence is emerging that the lifestyle choices and environmental conditions of parents and grandparents seem to have epigenetic consequences that can cascade down the generations. Problems such as obesity and schizophrenia may have some of their roots in epigenetic effects caused by diet, trauma and pollution, for example. The ultimate aim of scientists researching this is to observe and catalogue the epigenome – the millions of epigenetic switches that control our gene action. Scientists hope that this 'Epigenome Roadmap' will be the key to understanding the connections between disease and traits and our epigenetic markers.

In short, it's *really complicated*. The characteristics of an organism will depend on a whole host of factors: its genes, how its DNA strand happens to be coiled up within a cell nucleus, cell activity, the chemicals in the cells, the interactions between cells, and external conditions such as food, stress or pollution. All of which adds up to a far more nuanced situation than just 'It's in my genes.'

Let's use intelligence as an example. Studies of twins, adopted children and families indicate that a large part of intelligence is inherited. A lot of attention falls on a gene called *FNBP1L*, as well as a complex cohort of other genes. However, predicting intelligence is really not an exact science.

For a start, the environment is *very* influential here. Factors related to a child's home environment and parenting, education and availability of learning resources, and nutrition, among others, all contribute to intelligence. A person's environment and genes influence each other, and it can be challenging to tease apart the effects of the environment from those of genetics. For example, if a child's IQ is similar to that of his or her parents, is that similarity due to genetic factors passed down from parent to child or to shared environmental factors? It's most likely to be a combination of both. What's more, genes for traits like intelligence can come and go, depending on the environment. External pressures might bring some genes' advantages to the fore, or make them irrelevant.

Finally, how do you define intelligence anyway? The popular method has always been through IQ tests, which involve solving abstract puzzles and other mental challenges. But average human IQ rose about thirty points through the twentieth century, and our genes have hardly changed at all. Either we're getting smarter without genetic change, or IQ is more a measure of our culture's demands upon our brains.

It's likely that people are simply better equipped than ever to take the IQ test.

And talking of being well equipped, it's time for our second question. In the world according to *Gattaca*, scientists reckon they can determine a human's health, lifespan and personality by tweaking the genome. **Can we really use genetics to predict our destiny?**

## Born to rule

Do you think genetic engineering could have improved you, Michael?

Actually, yes. My genome doesn't make a protein that removes cholesterol from my bloodstream. So my high cholesterol could be fixed with genetic engineering.

It might be cheaper to fix that by eating fewer pork pies.

What? Give up my main source of pleasure? Not a chance... Anyway, what about you?

Well, science has yet to find a flaw in the Edwards genome. Which is not surprising, really, when you look at me.

You've just given me a new life goal: I am going to make sure I'm around to attend your funeral.

 Fair enough. I'll make sure there are pork pies at the wake.

In *Gattaca*, society is governed by genetic determinism. The idea is that, given access to someone's DNA, you will be able to predict the outcomes of their life. It doesn't matter what an individual does, they will always be governed by their genes – your destiny, even your date of death, is in your DNA. Bad with numbers? That's your genes. Scaly feet? That'll be your genes again. Depressed? Genes.

According to this way of thinking, there is no escape. And your genetic code trumps anything else. As Vincent points out, it doesn't matter how well he does in any test. 'My résumé is in my cells,' he says. If he doesn't have the right genetics, he won't be getting onto *Gattaca*'s space programme. Could this happen to us?

Well, it's certainly getting cheaper and cheaper to perform genetic screening – and it's becoming possible earlier and earlier in the life of a child. You can get the entire genome of an embryo sequenced for less than £1,000 these days. Don't worry about missing out, though: adults can get theirs sequenced, too.

The positives of such screening are pretty clear – as time goes on, our knowledge of the genome function will only increase, meaning that we will be able to tell more and more about medical conditions and other traits. This will allow

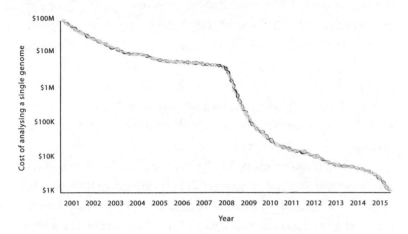

**The cost of analysing a genome is dropping rapidly.**

us to mitigate predispositions for certain diseases – and also to select embryos that don't have markers for serious conditions. But these genetic signatures have a dark side. There is grave potential for misuse if they fall into the wrong hands, or hands that don't quite understand the implications of the information.

It's not impossible to imagine a situation where genetic testing becomes a requirement imposed by insurance companies, for instance. You can be sure they would find your genetic information very interesting, when putting together your life-insurance package. If they consider your genetic risk factors to be high, say hello to a hike in your monthly payments. And then there's the workplace. Would a company take a dim view of hiring you, if they could access information showing that you are predisposed to certain health problems? (Hint: yes.)

Such situations are not just hypothetical. In 2012, a school in California made a decision to exclude a twelve-year-old student called Colman Chadam because he had genetic markers for cystic fibrosis. Markers are no guarantee of developing the disease, and little Colman didn't have it. However, kids who *do* have cystic fibrosis need to be kept apart, because they're particularly vulnerable to contagious infections carried by other sufferers. The school already had two other students with cystic fibrosis, and decided that this boy's genetics made it better to exclude him.

This is a good example of the problem with genetic determinism: just having a gene doesn't mean anything. Our growing realization that reading genes is not an exact science was prefigured in *Gattaca*. Uma Thurman's character, Irene, is genetically 'perfect', in theory. In reality, she shares a condition with the genetically 'imperfect' Vincent: a weak heart. 'Mine is already ten thousand beats overdue,' Vincent tells her.

She is astonished that that is even possible. Before meeting Vincent, Irene subscribes so wholly and unquestioningly to *Gattaca*'s extreme genetic determinism that she would, in the words of the film's writer/director Andrew Niccol, 'lie down and die at the allotted minute because she would feel guilty if she lived a minute longer than her profile prescribed'. This idea of genetic inheritance being equivalent to predestination is ultimately skewered in the film, as Vincent and Irene both realize they don't have to be trapped by their genome.

And neither do you. Which, thanks to decades of sloppy journalism, may come as a surprise. After a quick flick through the newspapers in recent years, you could be forgiven for thinking that we are now on our way to finding the 'gene for' everything.

With some characteristics, there is a bit of validity to this idea. Eye colour, for instance, is determined by (relatively) simple genetic equations. Dimples, blood type, a chin cleft, knuckle hair and whether your ear wax is dry or oily are just some of the things that are fairly predictable, based on your parents' characteristics. But other factors, such as height and skin colour, are part of the much bigger group of physical characteristics that come from a complex mix of genetic and environmental influences. This group includes a wide range of physiological factors that can lead to a shortened lifespan. Michael's 'familial hypercholesterolaemia', which makes his blood cholesterol higher than average, is one. Despite what Michael's whingeing suggests, that's not even an especially bad one: some forms of cancer and diabetes are more likely to manifest in people with certain combinations of genes.

And behavioural characteristics are even more complicated. Genetics can influence your personality, but they don't *determine* it – they don't set it in stone – because environment, upbringing and lifestyle have strong effects here, too.

A good example is the so-called warrior gene. Not long after *Gattaca*'s release, the media had a lot of fun with this one. The monoamine oxidase A (*MAOA*) gene was touted as something that makes people violent. But the research didn't really stand up. At best, it was a gross oversimplification; at worst, just plain wrong. However, that didn't stop a murderer in the US trying to avoid execution in 2005. He claimed that he did what he did because of a mutation in his *MAOA* gene. The judge turned down the appeal, because the defendant wasn't a slave to his genome – or at least not in this simple way.

That said, we can't deny there are genetically influenced tendencies in all of us. That makes it important to distinguish between behavioural traits and actual behaviour. Traits are basically very broad inclinations. Your genes, although rarely

## What's in a name?

Geneticists who discover a new gene get to name it. Here are some of our favourites:

**Swiss cheese**: This is found in mutant flies whose brains have Swiss cheese-like holes.

**Tigger**: A transposable 'jumping gene' that can move to different places in the genome.

**Cheap date**: Mutations in this gene give flies a heightened susceptibility to alcohol.

**Drop dead**: This is not a good gene to have: it causes sudden early death in adult fruit flies.

**Fruitless**: This causes a defect in which male fruit flies lose interest in females.

**INDY**: This gene makes mutant fruit flies live twice as long (the name is an acronym for 'I'm not dead yet').

**ARSE**: *Arylsulfatase E* gene just didn't cut it as a name.

**Sonic hedgehog**: It's playful, but not helpful. This gene is now known to play a role in a developmental disorder of the brain, and doctors tend not to use the name when explaining to parents about the mutation that has put their child's life in danger.

single genes, *do* influence these inclinations, such as a tendency to take risks.

Displayed behaviours are different, though. They are actions taken, in the moment, which will partly be influenced

by traits, but mostly by situation, environment and the interplay between the other traits that you possess. In evolutionary
terms, traits are selected for (via natural selection) over time,
but behaviours occur as the current, situation-sensitive
expression (or not) of traits.

For example, let's say Michael has a variant of a specific
gene that deals with dopamine-processing (it's called
*DRD4-7R*, if you're interested). This gene is associated with
risk-taking and novelty-seeking (and, interestingly, attention
deficit hyperactivity disorder, or ADHD). Risk-taking and
novelty-seeking are traits, not behaviours. But how do we see
these traits manifested in Michael's behaviour? Well, Michael
might drink a lot, or sleep around. Does that mean we've
found a gene for alcoholism? Or promiscuity? Of course not.
Equally, Michael might simply travel more, read more and
have more friends. This gene isn't somehow coding for all of
these various behaviours; what it's doing is producing a broad
inclination towards being open and curious. How Michael
actually behaves, given this trait, will be down to opportunity,
environment and his other traits.

It's not hard to see that if Rick had the same gene, he
wouldn't necessarily behave in the same way as Michael.
Rick's and Michael's blends of genes are different (for which
they are both equally thankful), and that means that there
will be nuance to their curiosity. It won't manifest in the
same way.*

We seem to be astonishingly bad at admitting this.
Fidelity, criminality, religious persuasion... you name it, it
is often claimed that the behaviour is 'in our genes'. There

---

* Rick: it's worth pointing out here that Michael may or may not have
the *DRD4-7R* gene, but he sure does like a drink.

is still no shortage of articles proclaiming that scientists have found the 'slut gene' (seriously) or the 'genes for intelligence', and so on. However, very few scientists now believe that there is a simple 'gene for…' formula. It's been fun to try and boil down complex characteristics and behaviours to single genes, but the reality is that this is far too reductive. It is, you could say, complete and utter bollocks. The (let's admit it, slightly disappointing) truth is that genes just don't work like that.

Another complicating factor when trying to predict outcomes from the analysis of someone's genetics is that the interaction between genes is so important. It can even mean that two wrongs make a right. In 2008, for example, a study was published that looked at the interplay between two 'bad' genes. One was a variant of a gene called *SERT*, which seems to make people more affected by negativity and is associated with depression. The other was a variant of a gene called *BNDF*, which is involved with maintaining and growing neurons. This bad version is less good at its job, and means that people with that gene tend to have trouble learning some things. But here's the good news; if you have both in the same genome, the *BNDF* variant means that you're rubbish at learning the negative lessons the *SERT* gene is pushing, and you're less likely to succumb to depression. In simple terms, the two 'bad' genes cancel each other out. Result!

This leads neatly onto our third question. Should we correct the 'bad' genes? **Should we use genetics to make the perfect human?**

# It's a long way to the top

More *Gattaca* trivia: do you remember the middle name of Jude Law's character, Jerome Morrow?

The 'genetically perfect' specimen who would be displaying inherited baldness, if they'd made the film a few years later?

Don't sneer, you bitter old man. Anyway, it's Eugene, from the Greek '*eugenes*', meaning well born.

Hence 'eugenics', the science of perfecting the human race.

Whatever happened to that?

I take it you dropped History pretty early?

In 1979, the year Rick was born, Josef Mengele died. Mengele was the infamous Nazi doctor who carried out horrific experiments on human subjects in the concentration camps. He was fascinated by genetics, and sought out identical twins for his most gruesome investigations, hoping to learn from them which characteristics were purely hereditary. Mengele's work was directed towards building an Aryan 'master race' of perfect humans, and destroying all 'inferior' races.

Eugenics pre-dates the Nazis, though. The idea had been around for ages – in Plato's *Republic* he wrote about pairing the good with the good and the bad with the bad, and destroying the offspring of the bad couplings, all in order to keep the 'flock… preserved in prime condition'.

The term 'eugenics' was first coined by Francis Galton in the late 1800s, as a catchy way of expressing 'the science of improving stock'.* By the early twentieth century, eugenics was very popular in Europe and America amongst 'Social Darwinists', who practised the sterilization of those with so-called 'undesirable' traits. The first sterilization law in the US was passed in Indiana in 1907, intended to give prisoners vasectomies and thus prevent the transmission of 'degenerate traits'. The US president at the time, Teddy Roosevelt, said that 'criminals should be sterilized and feeble-minded persons forbidden to leave offspring behind them'. By 1936, thirty-one states in America had some form of eugenics or sterilization law in place, and more than 60,000 people had been sterilized in the US by the time those laws were abolished. Some states are now proposing to give compensation to those who were treated.

Nazi Germany saw the most extreme eugenics programme. In 1933, the Reich government passed the Law for the Prevention of Genetically Diseased Offspring, which allowed the sterilization of anyone with hereditary physical or mental disabilities, including 'feeble-mindedness', depression, epilepsy, blindness, and so on.

Not satisfied with mere sterilization, in 1939 Hitler introduced 'mercy deaths' for the 'incurably sick'. By 1941, 70,000

---

* Galton was an eminent scientist and would probably be quite annoyed that you've never heard of him, but have definitely heard of his half-cousin, Charles Darwin.

German patients had been euthanized. In the next few years euthanasia was standard practice in Germany, and it's believed that 200,000 people were killed by the programme.

The spectre of the Nazi eugenics programme looms large over discussion of a 'new eugenics' made possible by the advanced genetic screening methods available today. Needless to say, we have since been extremely wary of any science that moves in this direction. But we do want to use our growing understanding of genetics to minimize pain and suffering, don't we? Guys?

*Gattaca*'s title card carries a chilling quote from psychiatrist Willard Gaylin: 'I not only think that we will tamper with Mother Nature, I think Mother wants us to.' Is that baseless self-justification, or a reasonable argument that evolution has now made us clever enough to intervene directly in the processes of Nature?

Two per cent of all babies – that's millions of babies every year – are born with disabilities caused by genetic defects. Millions more have gene variants that we believe predispose them to illness or disease. So why would you not look out for these things? Why would you not have the 'best' child you can? If we are able to vastly increase our children's chances of being healthy and the technology is cheap and easy, isn't there a moral obligation to do so?

It's certainly becoming possible. For a while now a heel-prick test has been done on newborn babies – just like the one in the film, though with a less extensive readout. It screens for a handful of genetic conditions, including sickle cell disease, cystic fibrosis and hypothyroidism. During IVF we screen embryos for genetic abnormalities before we implant one. So the scene where the geneticist is discussing with Vincent's parents which embryo to select – the embryo that will go on to be Vincent's brother, Anton – is already happening in fertility clinics. Though it seems dystopian in the film, maybe it's just sensible. We go deeper than health concerns, too: IVF clinics routinely allow women to select sperm donors by traits such as profession (in case you were wondering, doctors' sperm are the most popular).

And we are going even further: in an exciting *Gattaca*-style initiative, geneticists at Harvard University in Boston have started a programme where parents can sign up to have the entire genome of their baby sequenced. Cool, huh?

Apparently not. The Harvard doctors have been astonished at how few parents want to do this. Only around 7 per cent of the new parents approached by the programme agreed to get involved. It seems there's a bigger issue here, about how much we want to know about ourselves or our children, and how much we would change if we could. Are we ready for 'precision medicine'?

So far, we've only talked about weeding out disease by doing things like selecting the embryos that have the fewest flaws. In *Gattaca*, Vincent's parents choose the best possible embryo for his younger brother, but they are limited by their own gene pool – all the genetic material has to come from them. When they ask about leaving a few things to chance, the geneticist tells them that 'We have enough imperfection built in already. Your child doesn't need any more additional burdens.' But what if we had a way to introduce some different gene variants, or selectively turn some on or off? What if we could get rid of all those troublesome imperfections? Enter CRISPR, the gene-editing tool.

CRISPR stands for clustered regularly interspaced short palindromic repeats. In 2012 molecular biologists who were looking at how bacteria defend themselves against viruses found that they produce a bit of genetic material that is complementary (as in, will stick) to the genetic sequence of the attacking virus. This, along with a protein called Cas9, can then lock onto the viral DNA and disable it. Bacterium 1: virus 0.

Scientists have stolen this technique and are using it as a gene-editing tool. CRISPR/Cas9 acts like a very precise pair of molecule-cutting scissors. The CRISPR part is the guide, directing the Cas9 cutting tool to the right bit of DNA.

At the moment CRISPR can target, then disable or repair a gene, or insert something entirely new where it cuts. Gene Yeo, a biologist* at the University of California, compares it to a Swiss Army knife. At the moment, all it has are a blade and scissors, but Gene and his colleagues are bolting on proteins and chemicals that will transform the blades into multifunctional tools.

We can use CRISPR to tinker with the billions of chemical combinations in our DNA, turning off genes one at a time to see what effect that has. CRISPR can introduce specific mutations, to try and identify what causes illness or confers protection or other beneficial traits. It has already been used to tamper with the genes of plants and animals to create drought-resistant maize; goats with lovely long hair for cashmere; hornless cattle... The list goes on.

The first human trial has already been carried out – in China, where ethics isn't always a priority. A team extracted white blood cells from a patient suffering from otherwise untreatable lung cancer. They then modified the white blood cells using CRISPR: specifically by disabling a gene called *PD-1*, which usually stops cells calling out to the immune

* And poster boy for nominative determinism.

## Can we make a twelve-fingered pianist?

An intriguing scene in *Gattaca* has Vincent and Irene watch a twelve-fingered concert pianist perform a piece that can only be played by someone endowed with a genome that gives them a dozen digits. Is this possible? Definitely. But only if people don't get all squeamish about it.

The actress Gemma Arterton was born with six fingers on each hand, and had the extras removed shortly after birth. How boring of her! An Indian man called Devendra Suthar was born with fourteen toes and fourteen fingers and has kept them. Good lad! In 2016, a baby was born with fifteen fingers and sixteen toes (in China). A family in Brazil has fourteen members that have twelve fingers and toes, a clear indication of a genetic trait.

Having more than the normal complement of digits is down to a genetic abnormality known as polydactyly. It's surprisingly common – as many as 1 in 500 babies has some form of extra digit, though many are small and don't contain any bone.

Experiments on animals have shown that polydactyly can be induced by giving pregnant mothers certain chemicals (it's been done on rats, mice and – slightly weirdly – chameleons, but not humans). This suggests the genetic programming for finger formation can be disrupted. So – if we had no morals or ethics and we celebrated such diversity – yes, we could make a twelve-fingered pianist.

system for help; the hope is that the edited cells, which are allowed to multiply and are then re-injected into the patient's body, will gather at the cancer site and call in a strike from the immune system.

This is just one example of how we might go after disease with cells that are gene-edited outside the human body and then reintroduced, but it's hard to do on a large scale. Treating many conditions will require us to use gene-editing technology on the cells that are still in the body.

There are two pathways here. One is 'straightforward' gene therapy. This is the treatment of the 'somatic' cells, which are not involved in reproduction. Here, we can excise a gene, insert one or turn them on or off.

Editing somatic cells will not pass on changes to your children. That's not the case with the second technique: germline therapy. This involves the manipulation of the genome in sperm and egg cells or the early cells of an embryo. Here, things are different. These changes will be passed on to subsequent generations: they are a way of changing the human genome for ever.

Two teams (in China, obviously) have already tinkered with human embryos (naughty!). It was a move that prompted an international summit, at the end of 2015, on the ethics of CRISPR usage in humans. By the end of the summit, biologists had agreed a moratorium – a deliberate pause, effectively – on germline manipulation. However, it's over: Kathy Niakan of London's Francis Crick Institute now has permission to edit genes in embryos, as long as they are destroyed after seven days.

Which is why we need to talk about some of the darker sides of this technology. Tinkering around with our own genome risks undoing the gains of evolution – it could be that we introduce unforeseen problems that were edited out by the 'survival of the fittest' mechanism. We might suffer a catastrophic loss in genetic diversity. What's more, this kind of genetic manipulation will almost certainly be the province of the rich and powerful, so it's likely to exacerbate inequality, giving us a rich, genetically improved superclass who lord it over the poor. Exactly as *Gattaca* foretold, in fact.

So, to sum up: we are definitely more than just our genes, and *Gattaca*-style prediction is nonsense.

But we can, and will, use genetics to build the perfect human – whatever that is. I have to say, it all feels a bit scary to me.

You're not alone. Professor Jennifer Doudna, whose team was involved in the discovery of CRISPR, once had a nightmare where a man was sitting with his back to her and wanted to discuss the potential of her discovery. Who do you think the man was?

Josef Mengele?

Worse: Adolf Hitler.

So her conscience is obviously clear.

# 9

# Ex Machina

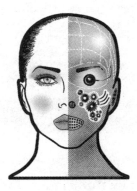

WHAT IS ARTIFICIAL INTELLIGENCE, AND WHAT CAN IT DO?

CAN A MACHINE EVER BE CONSCIOUS?

WILL WE SURPASS NATURAL HUMAN INTELLIGENCE?

This is a movie about a damaged orphan boy who comes up against a dangerous sociopath with seemingly unlimited powers.

I didn't realize we were doing *Harry Potter*.

> Very good. Did you know that Domhnall Gleeson, the actor who plays Caleb in *Ex Machina*, was once head boy at Hogwarts?

 In real life?

No. Do you understand what acting is?

 Did you see me in *Chalet Girl*?

*Ex Machina* is a rare thing indeed: an Oscar-nominated Hollywood success inspired by an academic tome. OK, so Murray Shanahan's *Embodiment and the Inner Life: Cognition and Consciousness in the Space of Possible Minds* isn't as dry as some academic books. But it isn't *Fifty Shades of Grey*, either.

Here's the plot: Nathan (Oscar Isaac) is a software guru who has used all the data on his search engine BlueBook to train an artificial intelligence. He has given this AI a series of embodied forms, the latest of which is Ava, played by Alicia Vikander. Nathan has brought one of his employees, Caleb (Domhnall Gleeson), to his remote lab/luxury playboy pad to see if he believes she has true intelligence.

Ava is an entirely lovely, sympathetic character. Until… *Oh no, look out, Domhnall!* Being head boy at Hogwarts has not prepared him for dealing with a flirty fembot. Not at all. Ava, it seems, resents being tested and starts to use Caleb as a part of her escape plan. Or is this all part of Nathan's elaborate ruse?

To get to grips with this film, the first question we have to deal with is a relatively straightforward one: **What is artificial intelligence, and what can it do?**

## Rise of the machines

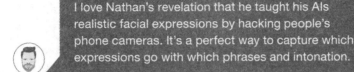

I love Nathan's revelation that he taught his AIs realistic facial expressions by hacking people's phone cameras. It's a perfect way to capture which expressions go with which phrases and intonation.

It wouldn't work. Because all you'd get is a close-up of someone's ear.

Not true. They could be using a hands-free set.

Well, then you would just get that smug 'Look at me, I don't need to hold a phone to my ear, I am from the future…' face.

I think you're being overly harsh. Have we hit on something deeper here? Does nobody ever call you?

Not even my mother.

You're first in the queue for a robot companion, aren't you?

Y ou are intelligent. Of course you are – you're reading this book. But could you ever say the same of a machine? The question of machine intelligence was ridiculed for decades after it was first raised. In 1948, for instance, Alan Turing, the pioneer of modern computing, wrote a research paper called 'Intelligent Machinery', a first foray into the way a computer might mimic the operations of a human brain. 'I propose to investigate the question as to whether it is possible for machinery to show intelligent behaviour,' Turing wrote. 'It is usually assumed without argument that it is not possible.' Turing's boss at the National Physical Laboratory in London – a man who gloried in the name Sir Charles Darwin (his grandfather, *the* Charles Darwin, never got a knighthood: ha!) – was unimpressed. It was a 'schoolboy essay', in Sir Charles's view, and was not to be published.

That didn't put Turing off. Two years later he published a paper called 'Computing Machinery and Intelligence', which contained a provocative question: Can Machines Think? The paper proposed finding the answer via what Turing called 'The Imitation Game' (sound familiar?), where a hidden computer holds a conversation, via some kind of communications technology, with a human. If the human can't tell it's talking to a computer, the machine can be regarded as having 'artificial intelligence'.

This, the 'Turing Test', is central to *Ex Machina*. Nathan, the multibillionaire creator of the robots, tells his lucky employee Caleb that we need a new kind of Turing Test, because artificial-intelligence research has advanced so far since Turing's day. The movie follows what is, essentially, the new test. And it's pretty terrifying.

Today's most successful AI starts with 'neural networks'. These mimic the composition of our brains, where small bic logical processors called neurons are connected to each other in a complex network. As we discovered when looking at *Planet of the Apes*, a neuron is a cell that reacts to an input signal by generating an output signal. The output depends on both the input and the particular characteristics or settings of the neuron.

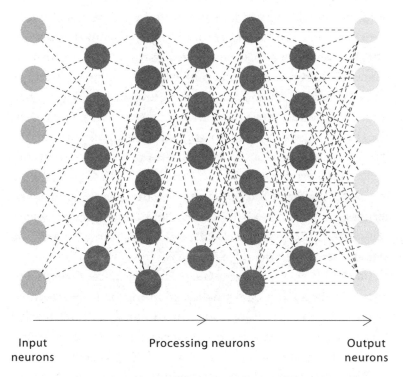

Input neurons      Processing neurons      Output neurons

**A neural network's input signals are fed through the neurons via connections that change as the machine learns how best to achieve its goals.**

And as we learned in our *Matrix* chapter, the human brain combines the complex interweaving of inputs and outputs with feedback from the sensory system, such as the eyes, ears, skin and pleasure centres. The result is that our experiences 'reinforce' certain pathways, changing the biochemical make-up of the neurons and the strength and quantity of connections. We call it learning.

Things are no different in a machine that learns. Artificial neurons are small, silicon-based components that process an input to give an output. The inputs and outputs are the connections between neurons, and between the neurons and the outside world (or a machine that the network is trying to control). The connections between the neurons can get stronger or weaker – that means, effectively, that some neurons trigger each other more easily, while others require a bigger electrical signal before they begin to process an input and generate an output. Finally, there are basic goals, such as winning a game of chess.

When Garry Kasparov, the world's best chess player, first lost to IBM's Deep Blue computer, he described it as a 'shattering experience'. He had played a lot of computers, but this was different. 'I could feel – I could smell – a new kind of intelligence across the table,' he said.

But here's the plot twist: Deep Blue wasn't actually intelligent. It didn't use any kind of adapting or learning from experience. It just employed a brute-force approach that could harness its blistering processor speed to run through all the possible moves and pin down its optimal strategy.

There's nothing clever in any of that. Not clever like AlphaGo is clever. This is the machine that should really shatter Kasparov. AlphaGo has learned how to play Go, a deceptively simple Asian board game that involves surrounding your opponent's counters with your own. AlphaGo is now better than the best human players. That is truly astonishing, because humans have to play Go partly through intuition. Even the best players can't always describe why they make the moves they do. Sometimes they just have to look at the board and follow a gut feeling about the right response. No one can program gut feeling into a machine. But it turns out you don't have to.

DeepMind, the creators of AlphaGo, started their journey with basic neural networks playing arcade games. They constructed a sophisticated machine that could operate the controls for the Atari game Space Invaders. It quickly mastered that, and moved on to Breakout, where you use a bouncing ball to break through a wall. Points come from destroying the bricks. The DeepMind team didn't tell their 'agent' how to play, though. They just gave it the goal of maximizing its score.

Before long, it had discovered hitherto-unknown tactics for getting the highest possible score with the least possible effort. Not that it knew that. It didn't know anything. It was simply trying to push the score up as fast as possible.

If it could only do one job – play Breakout – DeepMind's agent would be what is known as 'weak' artificial intelligence. Weak AI is highly specialized in what it can do. It's a bit like being brilliant at carpentry; it's a useful skill, but it's not much help if you are suddenly asked to do the company accounts. What we really want is 'strong' artificial intelligence: AI that can turn its robot hand to anything.

And that is exactly what DeepMind is aiming for, via AlphaGo. Go is not an easy game. The rules aren't that complex, but the best way to play involves analysing the 1,000,000,000, 000,000,000,000,000,000,000,000,000,000,000,000,000, 000,000,000,000,000,000,000,000,000,000,000,000, 000,000,000,000,000,000,000,000,000,000,000,000,000, 000,000,000,000,000,000,000,000,000,000,000,000,000, 000,000,000 possible positions. That's more than the number of atoms in the universe.

AlphaGo's creators fed their machine a few million human moves from a few million positions, and it became able to predict, from looking at a position, what a human would do. It only got it right 57 per cent of the time. But now it stood a chance of winning.

With the creators' next move, the human players ceased to stand a chance. AlphaGo was set up to play against itself many thousands of times until it had learned how to win from a vast number of positions. Not by pure brute force, like Deep Blue, but by supplementing searches with an agile, unprogrammed intuition. Based on its experience of millions of moves in the thousands of games of Go it had played, something deep in AlphaGo's neural network could take a look at a Go board and come up with a sensible, often match-winning move. If DeepMind's researchers asked it why it made that move, it couldn't tell them. If they took it apart, the answer wouldn't be there in its circuitry. Something akin to intuition had, effectively, arisen – emerged – *in silico* through experience.

DeepMind has big plans for the technology behind AlphaGo, or whatever they call the next incarnation of this intelligent agent. They want it to work on real-world problems, such as diagnosing disease or finding new pharmaceutical drugs

It won't be alone. We already have AI agents that are phenomenal at medical diagnosis, for instance. A company called Enlitic operates an AI that can look at medical images and detect lung-cancer nodules more accurately and more quickly than a team of expert radiologists. Then there's Google's self-driving car, of course. This has to be able to learn from successes and mistakes, make decisions based on signals it receives from the outside world and operate safely in

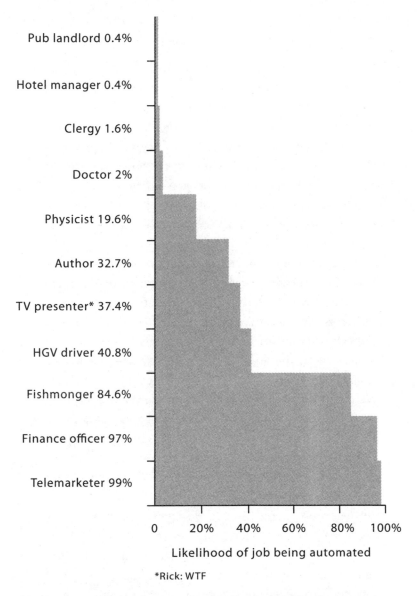

Pub landlord 0.4%

Hotel manager 0.4%

Clergy 1.6%

Doctor 2%

Physicist 19.6%

Author 32.7%

TV presenter* 37.4%

HGV driver 40.8%

Fishmonger 84.6%

Finance officer 97%

Telemarketer 99%

0    20%    40%    60%    80%    100%

Likelihood of job being automated

*Rick: WTF

Advances in robotics and artificial intelligence mean some professions have a significant chance of disappearing by 2035. (Data source: Michael Osborne and Carl Frey, Oxford University)

## Rooms with a view

In the movie, Caleb tells Ava about a famous thought experiment in artificial intelligence. It concerns a scientist called Mary, who has been raised in a room where everything is black or white; there are no colours. Mary knows all the physics theory about wavelengths and frequencies, and how the brain perceives them as different colours. But there is something missing from her understanding (and, similarly, a computer's) of colour: what it is like to see it. Even if we can understand everything about how the brain works, and replicate it *in silico*, will there be feeling?

The philosopher John Searle has made related observations in his 'Chinese Room' experiment. He imagined an AI that can read questions written in Chinese characters and use available resources to formulate an appropriate answer, also in Chinese. If the AI is in a closed room, the person asking the questions and receiving the answers might think there is a human Chinese speaker in the room. Searle's point is that the AI would pass the Turing Test, but wouldn't *understand* anything about those questions and answers. He showed this by imagining himself in the room with access to the instructions in the computer program, and all necessary resources to process the questions. He could then take in Chinese characters, and produce a correct output in Chinese characters. But, convincing as it would be to the questioner outside the room, Searle still wouldn't understand Chinese. And the original AI is the same, he argues. Just because you have the appearance of intelligence doesn't mean there's thinking, or a mind, or intention. We are too quick to mistake complex processing for intelligence, Searle reckons.

a chaotic, unpredictable environment. Artificial intelligence is the only way a machine could do this and, arguably, it does it as well as some human drivers.

More prosaic – but possibly more threatening – applications of AI are emerging, too. Real-time translation; journalism; speech recognition... the list is almost endless. None of these are going to create a sudden, radical shift in our way of life, just a slow, steady drip towards machines that can do much of what we have always considered to be tasks exclusive to human intelligence.

So how do we measure progress? Nathan is right: the actual outcome, the modern-day incarnation of AI, is so far beyond what Turing imagined that we have to find alternatives to the Turing Test. That's why Nathan wanted to know Caleb's reaction to Ava. Did he consider Ava a 'her' or an 'it'? Did he think she had emotions, feelings and goals? Did he think she was more person than machine? Which brings us to our second question: **Can a machine ever be conscious?**

## Sexy beast

 Would you have tried to help Ava escape?

I think so – I feel Caleb and I are quite similar.

 Except that you're not an expert programmer or even a half-decent hacker.

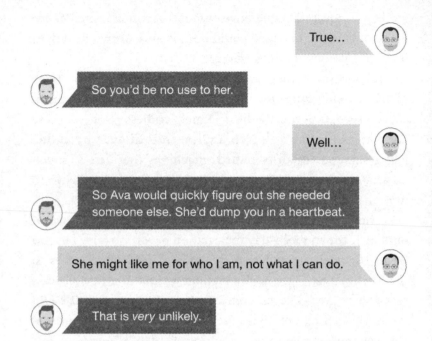

In *Ex Machina* it's almost like Nathan knew what was going to happen. 'One day the AIs are gonna look back on us the same way we look at fossil skeletons on the plains of Africa: an upright ape living in dust with crude language and tools, all set for extinction,' he tells Caleb. 'Don't feel bad for Ava, feel bad for yourself, man.'

Caleb is not quite so detached. In fact, he's confused about his unwanted and unexpected feelings for Ava. 'Did you program her to flirt with me?' he asks Nathan.

But Nathan didn't need to. He simply programmed Ava with a desire to stay alive, and the artificial intelligence that makes her who she is did the rest. That's why she worries about the test Caleb is performing on her. 'What will happen to me if I fail your test?' she asks him, railing against the injustice of being assessed on the vaguest of terms. 'Do you have someone who switches you off, if you don't perform as you should? Then why should I?'

It's a great question. In this light, 'I think, therefore I am' – René Descartes's famous justification of his conscious existence – seems a bit one-dimensional, doesn't it?

We saw in *Gattaca* that intelligence is a slippery concept. Consciousness is worse. Although there is no agreed definition, most accept that it has something to do with an internal self-aware state that experiences emotions and has goals beyond mere survival. The problem with that is that because all the signs are internal, no one can ever tell for sure if another organism is conscious.

So here's an interesting thought. Why should consciousness (whatever that is, exactly) emerge from a particular arrangement of carbon-based molecules, and not from a particular arrangement of silicon-based molecules that seems to be able to perform roughly the same set of tasks? In other words, why shouldn't Ava be as conscious as René?

Animals provide an interesting comparison here. Many researchers believe that much of the animal kingdom is conscious. The behaviour of octopuses (our favourites) and dogs certainly makes it very difficult to deny they exhibit consciousness. After all, Inky the Octopus – a former resident of the National Aquarium of New Zealand – showed conscious intent when he escaped from his tank and headed to the ocean in April 2016.* Inky's keeper had left the tank's lid slightly open (either by accident, or because Inky can

---

* Inky's motivation may have relevance to our look at *Planet of the Apes*. Was he intelligence-gathering as part of the octopus plot to take over from humans?

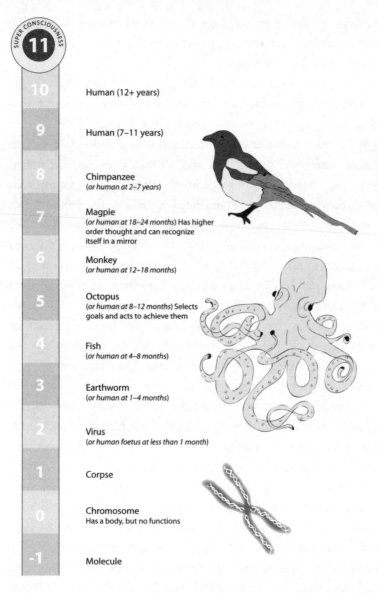

**SUPER CONSCIOUSNESS**

**11**

**10** Human (12+ years)

**9** Human (7–11 years)

**8** Chimpanzee
*(or human at 2–7 years)*

**7** Magpie
*(or human at 18–24 months)* Has higher
order thought and can recognize
itself in a mirror

**6** Monkey
*(or human at 12–18 months)*

**5** Octopus
*(or human at 8–12 months)* Selects
goals and acts to achieve them

**4** Fish
*(or human at 4–8 months)*

**3** Earthworm
*(or human at 1–4 months)*

**2** Virus
*(or human foetus at less than 1 month)*

**1** Corpse

**0** Chromosome
Has a body, but no functions

**-1** Molecule

Just how conscious are you? Spanish AI researcher Raúl
Arrabales Moreno and his colleagues created the ConsScale to
rank the kinds of consciousness different organisms exhibit.

manipulate the minds of others – we don't know) and the octopus took full advantage. In the night, Inky climbed out of the tank, disabled the CCTV system* and pulled himself through a fifty-metre drainpipe that opened onto the sea. Or so we think; he didn't leave a note explaining his Houdini-like powers.

Whatever actually happened, we surely have to concede that if animals are conscious, and we are conscious, it's hard to see why a sufficiently intelligent thinking machine – a radically souped-up, embodied version of AlphaGo – shouldn't also display signs of consciousness.

The big question is: How would we know for sure? This is the genius of Alex Garland's *Ex Machina*. If Ava can convince Caleb that she has feelings, desires and intent and is deserving of 'human rights', even though he knows Nathan made her, we sidestep the Turing Test question and have to face something much deeper. It's no longer 'Is it a machine or a human?' but 'Is this machine essentially the same as a human?'

Will we ever get to answer with a big fat Yes? Here, opinions are divided. For a start, even in the fictional world of *Ex Machina*, with a clear depiction of a highly sophisticated, embodied artificial intelligence, you can argue about whether Ava is conscious. Towards the end of the film, she gives a little smile when no one else is looking. That seems like a natural conscious response to a pleasurable experience, doesn't it? And what about her desire to avoid being deactivated – isn't that evidence of her Descartes-like existential angst? She seems to be suggesting that while she processes information about her existence, goals and purpose (or we could just say 'thinks'), she self-consciously exists. Is that not evidence of

* Not really.

## Sex and death in the age of robots

Alex Garland, the writer-director of *Ex Machina*, leaves us in little doubt that Nathan has made his robots female for a reason. Clearly, Nathan knows his creations inside out, as it were.

It's already becoming clear that in the real world, sex is a major motivation for the development of AI. And that could be a problem. If you buy – or rent – a robot for the purposes of having sex with it, some researchers worry it will diminish your respect for, and desire for interaction with, humans. That doesn't have any upsides, according to the Campaign Against Sex Robots. It's one thing to replace factory workers or taxi drivers with artificial intelligence. To replace human relationships, with all their subtlety and complexity, is quite another. Empathy will fall, the campaigners argue, and real people – women especially – will suffer.

And talking of suffering, the idea of having robots replace human military decision-makers is just as questionable. At the moment, we don't have any AIs that are authorized to use lethal force, but plenty of gung-ho military types are arguing that we should.

In some ways, you can see their point. The machines have lightning-fast information-processing and target-identification abilities. They can provide statistics on the likely outcomes of pressing the trigger. You could argue that they are better suited to modern warfare than humans. News reports of collateral damage resulting from human decision-making add fuel to their fire.

> The big question is whether you can ever have enough checks in place to make sure that mistakes don't happen. Currently, we keep a human in the loop. We don't trust AI to make life-and-death decisions over humans. And *Ex Machina* suggests that might be a good thing.

her inner life? Also, Ava is aware of her physical body and wants to adorn it with clothes, about which she seems to have feelings.

We can read all of this as evidence for personhood. But not everybody does – Alex Garland, the writer-director, and Murray Shanahan, whose book provided the film's inspiration, both have reservations about whether Ava should be considered conscious.

But, as we have noted, Garland and Shanahan can't prove it either way – no one can. The only person or thing you can definitively say is (or isn't) conscious is you. You alone hear your inner monologue. You are the only reliable source on what pain and love feel like. Everyone else might *tell* you they experience these things, but you can't ever verify that they're not just repeating some programmed spiel about feelings that is designed to fool you.

But is it even possible for such a 'philosophical zombie' to exist? Those who believe we can build conscious machines say no. If it has the capability to display all the traits of consciousness, it must have the capability to be conscious. Otherwise we're invoking some special 'essence' that you must be endowed with to be conscious. It's a bit like the ancient idea of the 'breath of life' that somehow animates biological organisms. It has to be better to assume consciousness is a property that emerges from certain forms of complex information-processing equipment than to imagine some special, mysterious stuff that turns biology into self-aware beings.

And if that is the case, we can make a conscious machine, albeit silicon- rather than carbon-based. All we need is more complexity in our engineering. Each neuron in the human brain takes inputs from around 10,000 others, and its output goes to 10,000 neurons, too. When the whole brain contains eighty-six billion neurons, it's a big task to reproduce the necessary complexity. But whaddaya know – some people are already trying to do it.

For starters, there's IBM's 'neuromorphic' chips. These are modelled on the brain of mammals. You start with around 6,000 transistors (these are digital switches, the building blocks of all computers) assembled to make a basic neuron. Then you connect hundreds of these neurons together in a way that creates more than a quarter of a million connections between them. IBM add on a memory module to manage information that needs to be stored, but it's essentially like a little tiny bit of a mammal's brain. Adding more and more of these modules is proving useful – the neuromorphic chip-based brain is capable of rudimentary learning already. As it gets bigger (a lot bigger, to be fair), it may have the potential to show signs of consciousness.

If that worries you, let us add to your concern. The whole thing is sponsored by the US military. The aim is to put these brains into drones that can then make their own decisions about what is interesting in a landscape, and maybe what to follow – or destroy. We already have drone-borne missiles capable of autonomously identifying targets and making the decision to open fire. So far, we haven't used them, but...

Rather less threatening is Pentti Haikonen's robot, XCR-1. It is a little box that would sit on the palm of your hand. It has wheels and eyes, a voice and a pincer-like grabber attached to its front end. As it moves around its environment, it talks to itself and responds to basic questions. It's cute. Which is what makes it so upsetting that Haikonen has a habit of whacking it every now and then.

The whack is to give the robot associations between what it's seeing and what it's feeling. 'Pain' disrupts its normal functioning, so it avoids things it associates with this sensation. So, in what is basically a dickish move, Haikonen gives it something green to look at, then slaps its pain sensor. XCR-1 learns that green is bad, and will then avoid contact with green things. It's pretty distressing to watch.

Haikonen does make up for it a little bit. He can also stroke the pleasure sensor, creating positive feelings towards what the robot is seeing.

Feelings? That's a very contentious word. But although Haikonen won't say so explicitly, XCR-1 could be described as having a sliver of consciousness. As with Ava, it's really up to you to decide. Go online, find Haikonen's YouTube videos and ask yourself if the age of conscious machines is already upon us. You might be surprised about the answer you're tempted to give.

So, how far can this go? It's time for our third question: **Will we ever surpass natural human intelligence?**

# A beautiful mind

Clearly there are limits to Ava's intelligence.

 What makes you say that?

When she went out into the world at the end of the film, did she take anything? Did she have a bag?

 No.

So where's her charger? With that brain, I give her a day at most before she shuts down.

 Good point. And she was planning to spend her first free day people-watching at a traffic intersection. Not exactly the apocalyptic *Rise of the Machines* scenario we're used to.

Less *Terminator*, more Terminate Her – before she dies of boredom.

 It's disgusting how pleased you are with that pun.

One of the most chilling moments in *Ex Machina* is when Ava whispers something in the ear of Kyoko, another of Nathan's creations. That is the point when you start to really worry for Nathan and Caleb. The robots are conspiring. They're ganging up on the humans. That can't end well.

It's not hard to imagine why. When we have created a human-level intelligence (let's not get distracted by issues of consciousness for now), we have to assume it will have an inclination towards self-improvement. After all, humans do, so it's likely to be included in the program that brought the AI into existence. We want machines to be smart, so we will help them to help themselves.

And what will a self-improving human-level artificial intelligence do? Well, it won't stop where it is. It won't think, 'Do you know what? I think we can all agree that this is about as clever as anyone, or anything, will ever need to be.' No, it will make itself a little bit cleverer than human-level intelligence. And then, within a few iterations, it will be a lot cleverer than any human that has ever lived. It will have the resources to make itself as clever as is physically possible. It will become super-intelligent.

Then its super-intelligence will probably cause it to make copies of itself – you know, just in case... It might even build in a bit of variety. And then it will start to enjoy interacting with these near-copies of itself. Before long, humans will be a tiny speck at the back of its vast mind. Or minds. Whatever. At that point, we're screwed.

This scenario is known as the 'technological singularity'. It doesn't actually have to be all doom and gloom, though. The apocalyptic scenario described above is only one possibility. Another is that we allow ourselves to fuse with the machines and then just go with it. Do you wanna build a cyborg?

## Survival of the fittest

In 1942, the science-fiction writer Isaac Asimov published the first coherent attempt at a set of laws for robots' decision-making:

1 A robot may not injure a human being or, through inaction, allow a human being to come to harm.

2 A robot must obey the orders given to it by human beings, except where such orders would conflict with the First Law.

3 A robot must protect its own existence as long as such protection does not conflict with the First or Second Laws.

But an AI is not a robot. Is it subject to the same rules? After all, it might consider itself not a servant or slave, but a being with rights – a being who doesn't understand why humans should be privileged.

It's an intriguing prospect. We could give ourselves implantable memory upgrades, silicon-enhanced neurons that fire a million times faster, augmented senses – all the good stuff. We would be superheroes, and ever more able to design upgrades and further enhancements. Could we get there? Will we get there? How? Here are the answers, in order:

Yes!

Maybe!

No one has a clue.

That third one's a let-down, isn't it? But the truth is, when you start to ask questions about how we will achieve human-level artificial intelligence – that is, go beyond simple extrapolations of what we have now – there is very little to say. Very little that's believable, that is.

One of the great visionaries of the technological singularity is a man called Ray Kurzweil. At the end of the 1970s, Kurzweil was creating reading machines for the blind. That laudable achievement led him on to developing a range of synthesizer keyboards with Stevie Wonder in the 1980s. He has since gone on to other things: he is, for instance, a Director of Engineering at Google. His website doesn't shy away from listing some of his accolades: the National Medal of Technology, twenty honorary doctorates, and honours from three US presidents are on the list. He has also been inducted into the National Inventors Hall of Fame.

If Kurzweil is to be believed, the singularity will happen in 2045. How does he know? Well, in 2005 he predicted that we will have a decent model of the brain by the mid-2020s. Then, he says, we will 'be able to create nonbiological systems that match human intelligence...' That will happen in 2029.

How? Through nanotechnology, which will also be a mature field in the 2020s. Or that's what he said in 2005. In 2017 he moved this milestone to the 2030s while, weirdly, not moving the date at which AI will reach human-level intelligence.

Piffling details aside, Kurzweil's vision is that we will then be able to 'rearrange matter and energy at the molecular level'. With nanotechnology in place, we will be able to create nanobots: blood-cell-sized robots will 'travel in the bloodstream destroying pathogens, removing debris, correcting DNA errors, and reversing ageing processes'. And making new brains, it seems. 'We'll ultimately be able to scan all the salient details of our brains from inside, using billions of nanobots in the capillaries. We can then back up the information. Using nanotechnology-based manufacturing, we could recreate your brain.'

And this is only the beginning of this vision. 'The nanobots will keep us healthy, provide full-immersion virtual reality from within the nervous system, provide direct brain-to-brain communication over the Internet, and otherwise greatly expand human intelligence... As we get to the 2030s, the nonbiological portion of our intelligence will predominate.'

It's quite a prospect, but not everyone shares Kurzweil's optimism. Former editor-in-chief of *Scientific American* John Rennie calls it 'slippery futurism'; Kurzweil's predictions 'come with so many loopholes that they border on the unfalsifiable,' he says. Paul Allen, co-founder of Microsoft, now runs a brain-research institute (among other things, such as helping to hunt for aliens) and reckons we are a long, long way off doing anything so exciting in this area.

Allen does concede that the basic workings of the brain can be known, in principle. But you need more than the static blueprint: you need the dynamics. How does the brain react and change? How do billions of parallel neuron interactions result in human consciousness and original thought?

Allen reckons this kind of knowledge is subject to something he calls a 'complexity brake'. In other words, it looks doable at first, and progress even looks quite promising. And then it starts getting harder and harder, the further you try to go.

If you want a good example of a complexity brake, take a look at the progress of nanotechnology. Since they were seriously proposed in the 1980s, the tiny gadgets that mimic biological machinery have been slow to appear. In 2013, Eric Drexler, whose 1986 book *Engines of Creation* laid out this future, admitted that no one had actually started building them yet.

Things are looking up a bit lately, though. Researchers at Harvard University have, for instance, created nanobots made of folded-up DNA strands that can deliver drugs inside a cockroach's body. When they encounter the right kind of biomolecule, the DNA strands will unfold, releasing the drug dose it has been carrying. That means it can target specific disease-causing chemicals or cells.

We're still a long way from repair nanobots coursing through human bodies, however. And to put another nail in the coffin of the singularity, some argue that you simply can't outdo human intelligence with an artificial brain, no matter how smart it is. Our brains have evolved alongside (well, on top of) our bodies, and they have been honed by millions of years of evolution. Maybe it's the brain-plus-body combo that is the true source of our intelligence? After all, there's a lot of information carried in our genome (as we discussed in the chapter on *Gattaca*). Perhaps our brains only seem smart because they have evolved to work in tandem with the rest of our biology.

In other words, maybe AI isn't going to destroy us for the foreseeable future. And maybe we can still steer its progress so that it helps us build a better world, not a scarier one.

> It's a big 'maybe'. What have we learned? Artificial intelligence is here, and its capabilities are growing fast. No one knows if machines will ever be conscious – because no one knows what 'conscious' means...

 And one day I might be super-intelligent.

Would you upload yourself to a computer, to live for ever?

Obviously. You?

I don't know. It seems like a strange existence. I quite like my body.

You're alone there, I'm afraid.

# 10

# Alien

WHAT DO ALIENS LOOK LIKE?

ARE WE ALONE?

DO WE REALLY WANT TO FIND ET?

In space, no one can hear you scream. Brilliant line, that.

Except that, technically, they can. In 2014, Voyager 1 picked up a sound carried by a pressure wave in matter that had been ejected by the Sun.

Oh. Well, *Alien* came out in 1979. Voyager was only two years into its mission then. They didn't know.

 Plus, 'In space, no one can hear you scream unless you're lucky enough to have your terror conveyed on a coronal mass ejection' didn't fit on the poster.

*Alien*'s tagline has made its way deep into our psyche, offering a pointed insight into just how scary this film is. The effects look a bit dated at times now, and the franchise has bloated way out of control, but the original bite remains.

The concept was developed from a script where little critters cause problems on a Second World War aircraft. And 'developed' is the word: now we have an eight-foot monster terrorizing the crew of a futuristic spaceship. One by one, it wipes them out, until only Sigourney Weaver's character, Ellen Ripley, is left. Director Ridley Scott actually wanted the alien to finish the job, ripping Ripley's head off, then flying the ship into the darkness. He was overruled, of course: sequels are so much easier and more lucrative when they contain a character from the original. And although the alien might conceivably be described as 'a bit of a character' by its family (especially when it's had too much to drink at weddings), it's not the most relatable of movie stars.

Ridley does a good Steven-Spielberg-in-*Jaws* job of heightening suspense by keeping the monster mostly hidden, but we do get a few fleeting shots of his alien creature. There's the nasty, phallic little hatchling that bursts from John Hurt's chest and scuttles off (suspiciously like it's being pulled along a wire). Then there are glimpses of the fully grown beast, dripping acidic saliva* from its vicious teeth, snarling with that grotesque secondary mouth on the tip of its tongue. Sketched originally by H. R. Giger, the surrealist artist and professional nightmare-inducer, it's one of the most iconic creations in cinema history. But how accurate is that likely to be? Oh look, there's our first question: **What do aliens really look like?**

## Heavenly creatures

Did you know that the inspiration for the alien busting out of John Hurt's body was the writer's own Crohn's disease?

I did not. I thought it was just an obvious horror trope.

Dan O'Bannon woke up one night with unbelievable pain in his stomach. He said it felt like something was trying to get out of his body.

That might be the best ever resolution of writer's block.

The creature in this movie is a long way from the 'little green men' that the alien-hunters of the 1950s thought

---

* Actually KY Jelly.

we might encounter. But are either of these life forms anywhere near the truth? There are certainly lots of options for what an alien could look like and, given what we know about the evolution of life on this planet, we can speculate about some of them in a fairly informed way. But because all we know about is the evolution of life on this planet, we are barely able to comprehend the other possibilities. We will inevitably end up pondering 'life as we know it', for the most part, because we don't really have any idea how to ponder life as we, er, don't.

Carbon is the building block of all life on Earth. It's a special kind of element: carbon forms nice long 'backbone' chains that give a wide range of possible add-ons. Also, it makes bonds with other elements that are stable, but breakable. Because of this, carbon does a great job of supporting life on this planet, along with its chemical allies: oxygen, hydrogen, nitrogen, phosphorus and sulphur.

But it doesn't *have* to be carbon. Silicon is often suggested as an alternative building block for another form of life because it shares many chemical properties with carbon. Ultimately, though, it doesn't appear to be as well suited to the (mammoth) task in hand. One major issue is that when carbon combines with oxygen, it forms a gas – carbon dioxide – whereas silicon oxidizes to form a solid, silicon dioxide. Oxidation is an important process in our biochemistry, and the formation of a solid presents a serious problem for an organism: how do you dispose of it? It's not impossible, but it's undeniably simpler to deal with gaseous products.

So let's go back to the example that we know. Carbon-based life emerged on Earth very quickly. We think that simple life – tiny, single-celled organisms called prokaryotes

– sprung into being around 3.8 billion years ago, only a few million years after our hot little rock had cooled down enough to allow it. They're still with us – as bacteria, for example.

If alien life halted at bacteria everywhere but Earth, that would be excruciatingly dull. But we have to acknowledge it as a possibility. In fact, there are a whole host of reasons why life might appear, but not progress all the way to interstellar travel. However, we're not here to dig too deeply into the disappointing outcomes. Especially when we can make sensible guesses about the eventual evolutionary outcomes, if life does get its act together. One big question is around the evolution of intelligence itself. Is it inevitable? As we discussed in our look at *Planet of the Apes*, we don't know for sure. But if we look at our own world we see that intelligence and problem-solving ability have evolved independently in a whole variety of creatures, from dolphins to humans to crows.

We believe our intelligence to be the most advanced, but interestingly we weren't the first. The first advanced intelligence was likely to have been our old friend the octopus. At the risk of seeming obsessed with octopuses (you've read the *Planet of the Apes* chapter, we presume? And *Ex Machina*?), they are an interesting model for a potential extraterrestrial life form. First, this creature has evolved in conditions that are totally – erm – alien to our own. The octopus diverged from our common ancestor, which would have been a sort of sexy aquatic worm with light-sensitive pigments on its skin, at least 500 million years ago. That means octopuses have been doing their own thing for a long while; they've been adapting to some challenges and pressures that are nothing like ours, and yet they have arrived at some similar solutions to certain other things. Their eyes are remarkably similar to human eyes, for instance, and they have a high-functioning brain

## The Great Filter

Maybe no one is around because it's rare for a species to reach a point where their spacecraft venture beyond their solar system. Are we unique? Is there a 'Great Filter' that generally stops the evolution of life before that point?

The Rare Earth Hypothesis says that we might just be special: our particular planet could be extraordinarily well set up for the evolution of life. Also, the emergence of life about 3.8 billion years ago might have been a fluke, in which case we might be the only life of any sort anywhere.

An arguably better bet for a Great Filter in our past is the staggering two billion years it took for the transition from the basic prokaryotes to the more complex eukaryotes. These have cells that contain a nucleus and other features called organelles that make complex chemical-processing possible. So far as we can tell, this step only happened once, by accident. Without this lucky break, life on Earth may never have got beyond basic bacteria. So it may be that other planets are teeming with life that is never achieving complex forms.

Another option is that there is no Great Filter – it just happens that we are among the first civilizations to evolve intelligence, and that we are on our way, at the same time as many others, to becoming super-advanced intelligences who will ultimately colonize their galaxies. However, we are a young planet, so why would it only be now that intelligences are evolving elsewhere?

Finally, there's the somewhat alarming proposition that the Great Filter might be ahead of us. It might be that civilizations reach a point of technological sophistication such that they will inevitably wipe themselves out. Something to look forward to.

and intelligence. These are examples of what is known as 'convergent evolution', where the same solutions to problems – like how to see – emerge independently. It also implies that intelligence is a very useful survival mechanism in a range of environments.

So with that in mind, let's start trying to put together an alien. Intelligence lets you predict and affect what's happening around you, and overcome problems. So we can assume our alien has evolved intelligence. We know that awareness of the environment is always useful, suggesting that any intelligent alien will have some analogue of our sensory detectors.

Eyes have evolved between fifty and a hundred times on Earth, independently and in different environments. Around 97 per cent of Earth's animals have eyes, so it seems very likely that our alien will also have them. Those eyes won't necessarily be adapted to see well in the same visible spectrum as our own, however; that will depend on the peak spectrum of their sun, and the tasks that evolution has adapted them towards. Many animals on Earth see in different ways from us. Cuttlefish, for instance, see polarization – the orientation of the electric and magnetic fields that make up light and other electromagnetic radiation – and use it to communicate. Vampire bats detect infrared radiation to 'see' the blood vessels on their prey.

Our alien will probably have two eyes, which seems to be the most popular strategy on Earth (sorry, spiders, but you're in a tiny minority). And they will likely be forward-facing, which gives stereoscopic vision and thus depth perception. This is extremely helpful, in terms of both catching food and avoiding becoming someone else's food.

It will need some sort of nose, although this needn't be sticking out (and it might sense electric fields as well as chemical smells, like the weird paddlefish). Also, some sort of ears and a mouth for gobbling up food are likely. The presence of teeth is not essential – just ask birds – but if they're eating alien 'plants' that may be fibrous, they're probable. As the alien in the film demonstrates, teeth also do a good job of scaring your prey. And then chewing them up.

In all likelihood, our alien will be symmetrical – that seems to be a feature of life, probably because it allows the instruction book for building the creature (DNA, or its equivalent) to be a bit more concise. Overall size and shape will largely depend upon the gravitational strength of the planet, and the corresponding atmospheric density. If gravity is strong and the atmosphere is thick, evolution might create some big 'flying' aliens that are able to use their planet's dense air to gain lift and to swoop about.

Our intelligent alien will have a big brain and almost certainly a protective casing for it. That could take the form of an exoskeleton, like the creature in the film. The drawback of exoskeletons is that they tend to collapse under their own weight if they get above a certain size, and they restrict growth. That might limit our alien's size, and therefore its brain capacity. So an internal skeleton, including a skull, is the more likely adaptation.

If we assume our intelligent alien has developed technology (how else is it going to find us, to use us as hosts?), it's going to need the ability to manipulate objects. That means something like our fingers: digits that can grip and rotate. These could, as depicted in plenty of sci-fi, be a kind of prehensile tentacle. Or they might just be good old-fashioned hands on the end of arms. A land-based alien will also need some means of loco-motion. It's hard to see past some form of legs, and if it's got digits on the end of one pair of limbs, the symmetry principle suggests there is certainly a chance that it will have them on the others.

But you don't want to use all your limbs for locomotion. If you have the ability to manipulate things in your environment (unless it's with a tentacle), it's better to keep some limbs free for doing that. So our alien could well be standing upright on just two legs.

This is all sounding very familiar. The logic seems to be leading us to something that is, well, quite humanoid in appearance. Is that just because we find it so hard to think beyond our own existence? Not according to University of Cambridge palaeontologist Simon Conway-Morris. He believes that given convergent evolution, and the tendency of evolutionary processes to end up at similar solutions to environmental problems, 'Darwinian evolution is really quite predictable.' When evolution and natural selection are in charge, he argues, common themes emerge. Hence his idea that, actually, a hominid shape is the best solution for a sentient creature on any world (or at least, a world similar to ours). So much so that Conway-Morris thinks aliens would be 'eerily similar' to us.

So that's one possibility – a humanoid. Seth Shostak, senior astronomer at the SETI Institute, has another thought. Our planet is a youngster, at a mere 4.5 billion years old. There will be planets that are twice as old. That means there could be intelligent life that has been evolving for far, far longer than it has on Earth. He looks at where humans are currently, in terms of machine intelligence, and puts forward the idea that an alien civilization will have reached a point where they simply ditch their organic ties altogether: a civilization that transitions to being purely technological intelligences, leaving behind 'the quaint paradigm of spongy brains sloshing in salt water', as he puts it. One big boon of this would be that they could tolerate extremely long-distance travel – and are therefore arguably more likely to get in touch, or pay us a visit.*

One thing is generally agreed – to everyone's great relief. We can rule out an alien that is parasitic on humans. Parasites evolve with their hosts, and given that aliens haven't been here on Earth evolving with us (so far as we know), that's not a possibility. That means we can sleep more easily. If aliens do make physical contact with us, they probably won't be bursting out of our chest cavities.

So we have a few possibilities: some very basic bacteria (boring), a humanoid alien (eerie) or a machine alien (terrifying). Now on to the really tricky bit: **Where are they?**

## All quiet on the western front

> It's gutting that I'm not going to see aliens in my lifetime.

> Well, maybe it's a good thing. I don't know that we humans would react very well to an alien encounter.

---

* While Seth thinks that's an exciting prospect, a visit from aliens isn't everyone's cup of tea. But we'll get to that…

OK, so let's imagine you find yourself face-to-face with a weird, ugly creature from another world. What do you do?

I don't have to imagine. We created a hit podcast together.

In *Alien*, the crew of the Nostromo don't realize until it's too late that they've been set up by an Earth-based alien-hunter who put the android Ash on board. Unfortunately for them – or most of them, anyway – alien obsessives are rarely rational types. Especially when you consider the odds of finding ET.

So far, we've been searching for decades and found nothing. A real damp squib. Whenever we discover something unusual that originates from outside our planet, there is a clamour of hopeful excitement and an enthusiastic attempt to attribute whatever phenomenon we have observed to extraterrestrial life. But we've always been disappointed. Where is everyone?

In 1961, the astronomer Frank Drake came up with an equation to try to answer that question.* It has seven terms, and once you plug in your values, it comes up with an estimate for the number of detectable alien civilizations in the universe. The Drake Equation, as it has come to be known, is all very well. But there is a problem – working out the values to attribute to each term. Here are those terms:

- The rate at which new stars are born
- The fraction of stars that have planets orbiting them

* $N = R^* \cdot f_p \cdot n_e \cdot f_l \cdot f_i \cdot f_c \cdot L$, obviously.

- The number of habitable planets per solar system
- The chance that life emerges on a habitable planet
- The chance that intelligent life develops
- The fraction of civilizations that will have detectable technology
- The length of time for which a civilization will survive and send out signals.

We've been trying to get a handle on these terms ever since Drake first flagged them up. We're doing pretty well on the first three, these days. By various means we've now discovered more than 3,000 exoplanets, and that has enabled astronomers to make better estimates.

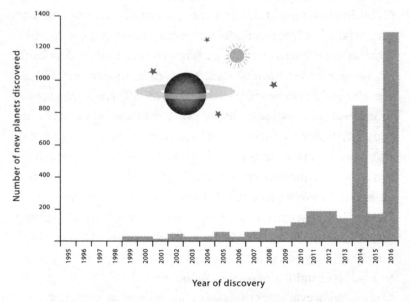

**We are getting good at discovering other worlds.**

We now think that 90 per cent of Sun-like stars might have exoplanets, and 20 per cent of those exoplanets are in a 'habitable zone' where the conditions that support life – at least life as we know it – should exist.

With the other terms, you pretty much have to take a guess (the chances of life and intelligence emerging are knowable in principle, but we don't know them yet). Entering the lowest, most pessimistic values, we can calculate that we are the only intelligent civilization in our galaxy, but there are probably 15,000 other intelligent civilizations in the observable universe. Going with some very optimistic values, we find there are more than 70,000 intelligent, communicating civilizations in our galaxy alone, and nearly eleven billion of them in the whole universe. That is a *lot* of aliens.

The other thing to consider is that Earth only formed 4.5 billion years ago. Given that we reckon the universe is 13.8 billion years old, it's reasonable to assume that many of the habitable planets we believe to be out there are much older. That means life will have been evolving for longer than it has on our junior planet. Consequently, as Seth Shostak points out, we'd expect some of those civilizations to be way, way more advanced than our own, possibly populated by super-intelligent cyborgs. And this suggests that, for various reasons including curiosity and the hunt for resources, these civilizations would seek to colonize other planets. Even travelling in craft at speeds that we can conceive of (say, a mere quarter of the speed of light), colonizing a whole galaxy like ours need only take some diligent aliens four or five million years. That might seems like a long time, but in cosmological terms it's an eye-blink. So, we'll ask again: Where the hell are they?

## How fast can we travel?

Voyager 1 is currently the fastest human-made object not in orbit – it's cruising along in interstellar space, having left our solar system, at about 61,500 km/h. That sounds fast, but Voyager would take nearly 80,000 years to reach Proxima Centauri, the closest star. If we were to send a crewed ship out on that journey, the crew that reached the star would be members of the 2,500th generation – 2,500 generations, in zero gravity, bombarded by radiation. Whisper it, but they might not even still be human…

Arguably the most exciting possibility for going faster is some sort of beam propulsion. It involves a huge, very thin light sail, being powered by focused energy beams (laser or microwave) generated back here on Earth. The Breakthrough Starshot project is aiming to use something like this to get an uncrewed nanocraft up and flying at 20 per cent of the speed of light. The hope is to launch a fleet of them 'within a generation' that could arrive at Alpha Centauri in a mere twenty years. Once there, the nanocraft will hopefully take some photos with their tiny cameras and stick them up on Facebook. Tag yourselves, aliens!

The design of the sail is obviously going to be pretty key. Some Harvard scientists have been looking at how to keep the sail at the optimal angle to pick up the propulsive beam, and have come up with a spherical structure. This would be self-correcting; if the craft wobbled left, the beam would naturally push it right. More importantly, these nanocraft will look like giant disco balls. If nothing else, the aliens will know that we're fun.

This is the question that the physicist Enrico Fermi asked in 1950, giving rise to the so-called Fermi Paradox. Fermi's point was really about the seeming impossibility of interstellar travel, but it's been interpreted as a reason to doubt the existence of alien intelligence. If there are so many aliens out there, surely we should have seen some evidence?

Maybe, maybe not. There are various explanations for why super-advanced civilizations might not have made themselves known to us yet. It could be that we're in a remote, desolate, 'rural' bit of the galaxy that the 'urban' aliens aren't that fussed about visiting. It might be that they've visited Earth thousands or millions or even billions of years ago and decided there's nothing worth scavenging. It might be that colonization just isn't of interest to a hyper-intelligent race. Perhaps they're homebodies who have found a utopian way of living in their own neck of the space-woods. They might just exist in a perfect virtual reality, rendering gallivanting around the galaxy decidedly unappealing. Maybe they're so advanced that we have no way of knowing that they're observing us, while operating a 'look but don't touch' policy: we might be an entertainment attraction, a curiosity – or a zoo. A more extreme version of that idea is that these aliens are developed so far beyond our conception that we can't even comprehend them. They could, in some way, be on Earth already, but we are just totally unaware.

Maybe, as in *Interstellar*, aliens live in the fifth dimension and we just don't know how to access their reality. Maybe we are like ants living in an anthill next to a ten-lane highway: both structures are significant, but differences in scale and speed of movement mean that the organisms using one structure might easily be blissfully ignorant of the other.

### Are we being abducted?

# NO.

A famous, but very iffy, 1992 poll suggested that 3.7 million Americans believe they've been abducted by aliens. CALM DOWN, AMERICANS!

The psychology of alien abductions is fascinating. First, the recollections of supposed abductees are often made under hypnosis. Hypnosis is not a reliable way of extracting 'hidden memories' – in fact it's been shown that it's pretty easy to induce false memories in a hypnotized subject, especially ones who are prone to suggestion. Second, many abductees appear to suffer from 'False Memory Syndrome', a tendency to recall words or items in memory tests that they've never seen before.

Sleep paralysis is also believed to play a big part in these narratives. Sufferers experience a temporary paralysis when falling asleep or waking up. It's a fairly well-understood phenomenon, and we know that when these people wake up, their terrified minds sometimes produce flashing lights and buzzing, the feeling of floating or the presence of figures (hello, aliens!). To be clear, these are just hallucinations. Most people suffering from this problem consider the effects to be a part of dreaming; others interpret them as evidence of alien tampering. The experience is subjectively real, but objectively… erm, a load of old tosh.

Research suggests that many people reporting abductions actively embrace the identity of 'alien abductee'. They seem to find it in some way comforting and psychologically helpful. It's like belonging to a creepy club.

Or maybe they haven't found us yet – and maybe we should be very, very grateful. Perhaps everything is so quiet because there are aggressive predator aliens out there, like the ones in the film, and the other intelligent civilizations know that and are lying low. Shit-scared and hiding, in other words. Which starts to make our activities – beaming signals skywards and sending craft out beyond our solar system – look a little foolish.

Stephen Hawking has confessed to being a little scaredy-cat on this point. He is worried that advanced alien races 'will be vastly more powerful and may not see us as any more valuable than bacteria'. That's as maybe, but the bad news is that the horse has already bolted. We've been broadcasting television and radio and radar for years, and these transmissions have leaked out into space. There's not much point in going quiet now.

The final explanation of absent aliens is, of course, the classic *Matrix* scenario – we're living in a simulation, and the programmers couldn't be bothered to code in any other intelligent beings. Maybe they just figured it was a waste of time, and also quite funny to watch us scratch our heads about it.

What, though, if there aren't any aliens? Now we're into a very scary suggestion. It might simply be that civilizations reach a level of technological sophistication where they will inevitably wipe themselves out. By engineering viruses that can't be controlled, or developing and deploying planet-purging nuclear weapons, or by creating technologies that blanket the planet in carbon dioxide and destroying the very conditions that once allowed them to thrive. It's not implausible, is it?

Really, though, we have *NO IDEA* about the absence of aliens. We're just speculating. Partly because it's extremely

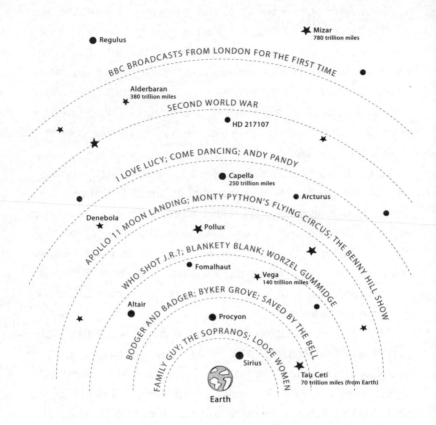

**Many of our TV signals have now reached other star systems.**

hard for us to get our heads around the possibility that we
might be an inferior species. We've had no experience of it
on Earth – we are a big fish in a small pond. Which makes it
entirely possible that we're really not in the best shape to deal
with an alien discovery, should it happen. And that's our final
question: **Do we really want to find ET?**

# Hello from the other side

Did you know there's a SETI post-detection committee that will take the lead in responding to an alien signal?

Of course I did. I've even interviewed its chair, Paul Davies.

Howzeee! And does it represent all of humanity?

If, by 'all of humanity', you mean a swathe of white people from Europe, America and Australia, plus one bloke from India… then yes.

No one from China?

No.

Don't the Chinese have the biggest and best radio telescope in the world?

Yes.

Have these people learned nothing from their childhood? You *never* exclude the kid with the best toys.

It must be nice to have nothing to lose. Ash, the crew's android, doesn't have that pesky biological imperative where you try to survive at all costs. He has orders from on high: if the mission encounters alien life, the priority is to capture it and bring it home alive. That's why he tries to convince his shipmates that this is the best plan, and why the crew's safety is not high on his list of priorities. However, when he's about to be taken permanently offline, he does offer his sympathy to Ripley about her chances against this 'perfect organism'. Nice robot. Ash has a point, though. Do we really want to find aliens, with all the dreadful possibilities that might ensue?

Maybe we haven't thought it through, but the evidence suggests that yes, we really do want to. We long for that moment depicted in *Alien* – intercepting 'a transmission of unknown origin' – and we've been looking forward to it for many years now. In fact, after decades of the Search for Extra-Terrestrial Intelligence (SETI), some enthusiasts are getting impatient with the passive approach of just looking for alien transmissions to arrive. They are pushing for a more 'active SETI' and say that we should be beaming signals directly towards promising locations – exoplanets in habitable zones, for example. We should be shouting 'hello', in other words.

Is that a good idea? Scientists just don't agree about whether there is genuine danger in broadcasting to exoplanets. Astrophysicist Neil deGrasse Tyson has pointed out that we don't give our address to members of our own species whom we don't know. 'So,' he says, 'the urge to give our home address to aliens? That's audacious.' It's a fair point. We have no way of knowing how our invitation will be taken. It might even be regarded as a provocation. Stephen Hawking famously compared a potential visit from aliens to being like the arrival of Columbus in America – a visit that worked out pretty badly for the original residents.

Others counter that if we or our resources were of interest, and aliens were problematic, they could have found us and plundered the Earth millions of years ago. The fact that they haven't is quite encouraging. And locating us is unlikely to be a problem for an advanced civilization, whether or not we broadcast a cheery hello. Alien astronomers could have detected the oxygen in our atmosphere (as we're trying to do with theirs) for the last half-billion years. And, as we said, we've been leaking radio, TV and radar for years. On that point, Seth Shostak, director of SETI, has said that it might be a good idea to send intentional messages, rather than leaving it to chance. Otherwise some aliens might tune into an old TV broadcast and get totally the wrong idea about us as a race. We don't want them judging us on past episodes of *Everybody Loves Raymond*, do we?

We have done some intentional messaging, in fact. We put something on the Voyager probes in the 1970s. Back in 2008, NASA sent a Beatles track – 'Across the Universe', chosen by someone with a very literal mind – in the direction of the North Star, Polaris, which is 431 light years away. Quite what the point of that was is anyone's guess. And what if the aliens have always been Stones fans, as well as being warmongering lunatics hell-bent on galactic domination? It would be a real shame if the Beatles ultimately proved to be the trigger for the demise of the human race.

Possible intergalactic conflict is not the only problem we'll meet when we come face-to-face (or face-to-proboscis) with aliens. Thanks to our enormous geographical separation, our respective civilizations are unlikely to have much in common.

## Updating the Golden Record

The Golden Record – humanity's account of itself, sent into space on the Voyager probes – is embarrassingly out of date. Luckily, we've got a new set of things to send out into space to make aliens aware of our greatness...

- A copy of this book, to give the aliens an overview of our scientific understanding (but also in the hope that they will order a few copies)
- The full range of 'Kimojis' (Kim Kardashian's emojis), as a simple way into human communication
- The Crazy Frog song, with a note explaining that this was a real low point
- The human genome as a how-to-make-your-own-human manual
- Some freeze-dried cheese, to make us seem less invasion-worthy
- A picture of Michael Fassbender, naked (just to show that we aren't to be messed with)
- A TV set to decode our broadcasts
- A jar of Marmite (to really confuse them).

And with beamed messages taking years (possibly thousands of years) to travel from one solar system to another, there won't be any snappy repartee, let alone a meeting of minds. Conversation is going to be stilted, with meaningful communication likely to be difficult, bordering on impossible.

Which is why it might make sense for us – or the aliens – to send probes with some form of artificial intelligence on board (maybe something not quite as shifty as Ash) that can analyse and learn the language, then communicate directly, asking and answering the questions that matter.

Using an AI makes sense for another reason, too. Carl Sagan once suggested that alien thought processes may well be running at very different speeds from ours. Much faster, or much slower. It could be that aliens are sending out 'hello' signals towards Earth, but the hello is over in a nanosecond, or drawn out over fifty Earth years. Either way, it would be hard to understand. And very difficult to start a meaningful dialogue where one party isn't bored out of its mind waiting for the response of the other. An AI might be able to cope much better than a human.

That said, our AIs are still built to our specifications and might not be able to deal with alien communications. Alien brains may have totally different architecture from ours. Is it reasonable to expect that we, or anything we create, can realistically understand alien cognition? We have a pretty bad record of communicating with other intelligent species on our own planet. Sometimes Rick and Michael struggle to find a common language. What's more, the aliens may have totally different values and beliefs. They may interpret things in a way that we wouldn't expect; most worryingly, they might see our 'friendliness' as aggression, or vice versa. We cannot assume they will be anything like us. In other words, it's all very risky.

For all the naysaying, there are plenty of good things that might come out of contact with an alien race. Optimists such as the psychologist Steven Pinker believe that human civilization has become more, er, civil with time, with fewer wars and better living conditions for the majority of people. So it's possible that an alien civilization more advanced than our own will be friendlier and more community-minded. In which case they might be able to teach us more about life, the universe and – well – everything. They might give us new technologies that end all human suffering. They might even be a good laugh. And who wouldn't want to see some alien art, or listen to the alien music charts, or even do a Captain Kirk and get off with an alien?

Let's face it, this whole endeavour – this whole book, in fact – derives from our irrepressible human curiosity. Seth Shostak, chief astronomer at SETI, thinks that our search for aliens, and for scientific answers, is driven by the same drive that has always led us to explore. This, he says, is a good thing for any society. Contact with aliens would allow us to understand more about ourselves and the cosmos. We might find out how much of the human experience is unique to us, and how much is universal. We could learn whether things like maths and science are fundamental, or just Earthly constructs. Perhaps some alien input would turn our understanding of ethics and morality on its head. So, yes, maybe it is a bad idea. But we should do it anyway. What's life for, if not taking risks?

 So aliens are probably a lot like us...

But they're a long way away, and that might be a very good thing.

I don't know why you're so pessimistic. Alien-hunting is humanity's greatest adventure.

Until we realize it's humanity's last adventure.

 Killjoy!

# Acknowledgements

Who do you want to thank first? Don't say yourself.

 Oh. I need to have a think then.

I'll help you out. Everyone at Radio Wolfgang, but especially our producers Max Sanderson and Hana Walker-Brown. Ivor 'Slayer' Manley, for his patience with your terrible mic technique. Cormac McAuliffe for being unfailingly clever. Colm Roche for running Wolfgang and occasionally buying us meals. And, of course, the icon that is George 'Larry's son' Lamb, for pairing us off, then running away...

 Max helped research and fine-tune the book, too. I've got another one. I'd like to thank Emer, for tolerating my dark moods whilst writing this and occasionally bollocking me for whining.

And, in general, for tolerating you as a husband. The woman is a hero. Oh yes, my wife: I'd like to thank Phillippa for her social utility.

Then there's all the experts who have cast their eyes over this and assured us that we haven't made any gross scientific errors: Murray Shanahan, Ronald Mallett, Simon Conway-Morris, Lewis Dartnell, Johnjoe McFadden, David Eagleman, Anjan Bhullar, Tracy Kivell and David Tong.

Yeah, if you spot a mistake, blame one of them.

Or you could blame the editorial staff at Atlantic Books, especially Mike Harpley, who steered this whole thing.

And it's my agent, Patrick Walsh, who got Atlantic on board. So he should surely shoulder some of the blame?

Seems fair. I'd like to thank Caroline Ridley, my agent, for having the good sense to just let me get on with it. That bodes well for my future projects. Which will not involve you, Brooks.

You're going to block me on all social media, now we've finished this, aren't you?

Already done.

# Index